T0263915

A Practical Approach to Water Conservation for Commercial and Industrial Facilities

A Practical Approach to Water Conservation for Commercial and Industrial Facilities

Mohan Seneviratne

Queensland Water Commission, Australia

AMSTERDAM • BOSTON • HEIDELBERG • LONDON
NEW YORK • OXFORD • PARIS • SAN DIEGO
SAN FRANCISCO • SINGAPORE • SYDNEY • TOKYO

Butterworth-Heinemann is an imprint of Elsevier

Butterworth-Heinemann is an imprint of Elsevier
Linacre House, Jordan Hill, Oxford OX2 8DP, UK
30 Corporate Drive, Suite 400, Burlington, MA 01803, USA

First edition 2007

British Library Cataloguing in Publication Data
A catalogue record for this book is available from the British Library

Library of Congress Cataloging-in-Publication Data
A catalog record for this book is available from the Library of Congress

ISBN: 978-1-85-617489-3

For information on all Butterworth-Heinemann publications
visit our web site at books.elsevier.com

Printed and bound in the United Kingdom
Transferred to Digital Printing, 2010

Dedicated to my parents Barbara and Yase

Contents

Foreword

That the planet Earth faces a looming water crisis is now beyond doubt. In many countries the great rivers no longer flow to the sea. Water is likely to increasingly become a source of conflict between regions and nations. Many countries and regions are facing the worst recorded droughts, and there is an increasing indication that climate change will have serious implications for water security.

This is one of the reasons why this book is so important. Water conservation is the quiet achiever for our sustainable water future. While there is a great emphasis on, and investment in, increasing supply, through more and more expensive and energy intensive means, it still remains the case that the cheapest, quickest and least risky 'source' of water for urban use lies in the savings that are possible in the millions of houses, offices, shops and factories in our cities and towns. Experience tells us that these savings are real, and can be tapped at low cost in all countries, regardless of their level of development. We must tap these savings, in order to reduce the cost of providing water services, to improve social outcomes and to capture the environmental benefits of reducing the extraction of water from our natural systems and reducing the discharge of wastewater to waterways and reducing the greenhouse gas emissions associated with pumping, treating and heating water.

This book focuses on the business sector, the industrial and commercial water users. This is an important area in the water conservation world, for many reasons. First, there are significant savings available, through both improved technology of water use, that is, more efficient equipment and better design of industrial processes, as well as through improved water using practices by staff. The involvement of management and staff in the collaborative development of solutions to improved water use is a key success strategy for obtaining savings in the business sector. Secondly, the savings can be achieved in large 'lumps' due to the high water use of individual customers. There are also economies of scale of rolling out programmes that target, say, all educational institutions, or all cooling towers – where there are homogeneous problems and solutions.

This book is important and timely, and it will provide an invaluable resource for managers, decision makers, utility staff and technical staff. It will help to build capacity in the emerging water conservation industry and to 'grow the market' of water saving. It provides a detailed description of how to obtain these savings. The author can rightly claim expertise in this area, as the foundation Manager of the Stockholm Water Prize winning programme, the largest such program in Australia, the Sydney Water Corporation Every Drop Counts Business Program.

This book will make a substantial contribution to the field. This means it will also make a major contribution to the urgent task of reducing water scarcity. And for that, it is worth its weight in gold.

Professor Stuart White
Director, Institute for Sustainable Futures
University of Technology, Sydney

This is an important book. It explains the difficulties we are facing in Sydney and elsewhere with climate change and related lower rainfall and higher temperatures. There is nothing new in this message but this book shows what can be done by business, industry and government organisations to use less water.

It is a practical guide based on the initial work and experiences that Mohan Seneviratne has implemented during his time at Sydney Water. Whether you are in the business of hospitality, running hospitals, servicing commercial office blocks, processing food, oil refining or operating a commercial laundry, this book is for you.

On a simple level we can all save significant amounts of potable water by being careful in our use. But what this work demonstrates is that a step-by-step process of tackling the issue of water conservation leads systematically to results that are inspiring. I take great pride in recommending this book. It is certainly one for the present day.

Kerry Schott
Managing Director
Sydney Water Corporation

About the Author

Mohan Seneviratne has over 26 years of water management, water treatment and water conservation experience. He received his MSc (Hons) in Chemical Engineering from the Moscow Institute of Petrochemical and Gas Industry, USSR in 1980. He has post-graduate qualifications in business management, applied finance and investment.

Mohan started his career in Sri Lanka commissioning ammonia and urea plants. From 1983–1999, he worked in Sri Lanka, the Middle East and Asia Pacific regions in technical consulting and marketing roles with specialisation in water treatment and water reuse applications in oil refineries, food processing, pulp and paper and thermal power plants.

Since 2001, Mohan has led Sydney Water's flagship demand management programme for the non-residential sector – Every Drop Counts Business Program. Over 350 of Sydney's high water using industries, government and commercial businesses are currently saving over 28 million litres of water every day. In recognition of these achievements the Stockholm Water Foundation in cooperation with Royal Swedish Academy of Engineering Sciences and the World Business Council for Sustainable Development awarded the 2006 Stockholm Industry Award to Sydney Water.

Mohan has written numerous articles for professional journals and presented papers at numerous conferences on water conservation, water efficiency and water reuse.

Acknowledgement

The current drought experienced in many parts of Australia and the world has been a wake-up call for many that the world's exploitable water is a finite resource and needs to be used wisely.

Whilst there are many books published on the need to save water, they primarily focus on residential applications. Industry uses vast amounts of water but very few books provide sufficient detail on how to save water in a practical sense. This book is an attempt to address that need.

This book draws on experience gained over 26 years in industrial water treatment, water management and water conservation spanning a wide range of industries. The last 6 years as Sydney Water's Program Manager for the Every Drop Counts Business Program, I was able to conceptualise and implement many of the concepts and principles of sound water management in the business sector. The book is organised to first give the reader a sound understanding of water conservation principles and then from Chapter 10 it focuses on five common industries. The principles and learnings expressed in the chapters could be used in other industries.

The book could not have been written without the help of a group of people who provided encouragement at critical points. I would particularly like to thank Anne Stuart of the Queensland Department of Natural Resources, my former colleagues at Sydney Water namely Steven Meleca, Sarah Baulch, Edward Maher, David O'Connell, Parameshwaran and my former manager John Nieuwland for reviewing chapters of the book and for making helpful suggestions.

In addition, I would like to thank the following for their insightful comments – Roger Horwood of Energetics Pty Ltd. for his feedback on Chapter 3 ('Saving Water Step by Step'); Bruce Smith, formerly Manager of the Hydraulic and Water Savings Branch, NSW Department of Commerce, and Rob Quinn, Managing Director of National Project Consultants, provided advice on Chapter 4 ('Flow Monitoring'); Michael Belstedt, Managing Director of Minus 40 Pty Ltd for his comments on Chapter 6 ('Alternatives to Cooling Towers'); John Koeller of the California Urban Water Conservation Council and Andy Gunasekare, Director of Engineering, the InterContinental Hotel Sydney for issues relating to Chapter 10 ('Hospitality') and

Dr Lisa Szabo, Chief Scientist NSW Food Authority and Sanath Nanayakkara, Quality Manager, Primo Smallgoods Pty Ltd. for comments on Chapter 13 ('Food Processing and Beverage Sector'). Finally but not the least my publisher Elsevier for having faith in the project and my ability to complete the manuscript and Padma Narayan of Integra for her skillful editing of the manuscript.

I also owe a debt of gratitude to my former employer Sydney Water for permitting me to publish extracts, photographs and other materials from publications such as *The Conserver*, Fact Sheets and Best Practice Guidelines. I am also indebted to the many suppliers and organisations such as the World Business Council for Sustainable Development, the Australian Food and Grocery Council and organizations and companies that permitted me to use their products, published materials, photographs and charts as examples cited in the book.

I am particularly grateful to Dr Kerry Schott, Managing Director of Sydney Water and Professor Stuart White of the Institute of Sustainable Futures, University of Technology, Sydney for writing the forewords for the book.

Finally but not the least I would like to thank my wife and two daughters for their support and patience.

Chapter 1

Water Conservation – A Priority for Business

> *Not a single drop of water received from rain should be allowed to escape into the sea without being utilised for human benefit –*
> *King Parakrama Bahu the Great of Sri Lanka (1153–1186)*

1.1 Introduction

Water is life. We recognise the value of water and its role in our day-to-day activities. Religions have recognised the role water plays in our well being. In developed societies, due to past investments in water infrastructure we have come to expect that water will be available 365 days of the year. We have being brought up with the notion that as long as we pay for it we have the right to consume as much as we want. Since there is no substitute to water, water prices have not reflected its intrinsic value and traditionally is subsidised. Consequently water is cheap relative to other resource costs. This has led to global fresh water consumption to rise faster than it is replenished. Between 1990 and 1995, fresh water consumption rose more than twice the rate of population growth. According to a report released by the International Water Management Institute (IWMI) at the Stockholm World Water Conference in 2006, a third of the world's population (roughly 2 billion people) is facing water scarcity now, not in 2025 as earlier predictions forecasted [1]. Water scarcity is not only a third-world problem. In recent years water scarcity have affected developed countries too. For example, in Australia the one in a hundred year drought has made water a political issue. It has highlighted the competing needs of agriculture, the low water prices enjoyed by the farming community which has led to wasteful practices and the need to supply the urban population with water where the majority of the populations live as well as the challenge to maintain environmental flows in the rivers. The world's water is in a crisis. But it is more a crisis of management of water rather than a water crisis. Therein lies the *triple* paradox of water. As the World Business Council for Sustainable Development (WBCSD) succinctly puts it: ***It is cheap, scarce and wasted*** [2].

Water quality is also decreasing and so far has not made the headlines. This alone could bring about a water crisis according to the 2006 Stockholm Water Laureate Professor Asit Biswas [3]. For an example, the rapid industrialisation of China and India will contribute to severe degradation of water quality in those countries if preventive measures are not taken.

Protecting the available water resources is therefore *our shared responsibility*. Business is part of the solution from the supply side as well as from the demand side. This chapter presents nine compelling arguments on why reducing water usage makes good business sense.

1.2   Global Water Resources Availability

From a global perspective, only 35 million km^3 (equal to 3.0%) of the world's water is fresh. Ninety seven percent is seawater and not readily available for human consumption. Of the 3.0%, permanently frozen in the Arctic and Antarctica, groundwater, swamp and permafrost constitute 2.5%. So that leaves only **0.5%** (equal to 105 000 km^3) in rivers and lakes to meet the needs of humans and the requirements of the planet's fresh water ecosystems. Figure 1.1 graphically shows the available global water resources.

Table 1.1 shows that 98% of the world's fresh water (0.5% of the total) is in aquifers.

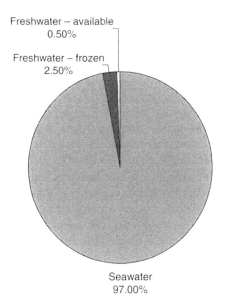

Freshwater – available
0.50%

Freshwater – frozen
2.50%

Seawater
97.00%

Figure 1.1   Global water resources [4]

Courtesy of the World Business Council for Sustainable Development – *Facts and Trends*. Geneva, Switzerland. August 2005.

Table 1.1 Where is this 0.5% of fresh water?

Water resource	km^3	Million acre-ft	Number of Olympic-sized swimming pools* ($\times 10^6$)	Percentage
Aquifers	10,000,000	8,107,013	4,000,000	97.9%
Rainfall on land (net of rainfall after accounting for evaporation)	119,000	96,473	47,600	1.2%
Natural lakes	91,000	73,774	36,400	0.89%
Man-made storage facilities	5,000	4,054	2,000	0.05%
Rivers	2,120	1,719	848	0.02%
Total	10,217,120	8,283,032	4,086,848	100%

* 1 Olympic-sized swimming pool is assumed to hold 2500 m^3 of water.
Adapted from the World Business Council for Sustainable Development – *Facts and Trends*. Geneva Switzerland. August 2005.

Case Study: World's Largest Aquifer Going Dry [5]

The world's largest aquifer is the Ogallala aquifer in the United States. It supplies water for irrigation to one-third of the United States crops and provides drinking water to Colorado, Kansas, Nebraska, New Mexico, Oklahoma, South Dakota, Texas and Wyoming. In other words it contains enough water to cover the entire United States to a depth of one-and-a-half feet. Even this aquifer is predicted to run dry in two decades due to over abstraction of water. Nebraska, Kansas and Texas were pumping 88% of all the Ogallala water between them, a massive 20 969 million m^3/yr (17 million acre ft/yr in 1991) more than the Colorado river.

Global demand for water needs to

- satisfy human needs for safe drinking water and proper sanitation
- expand agricultural production to meet population growth
- meet business needs to provide more goods and services for a growing population and
- minimise the impact of climate change on water resources.

1.3 Human Need for Safe Drinking Water and Proper Sanitation

The world's increasing water demands are driven by an increase in global population and urbanisation. The world's population is expected to increase from approximately 6 billion in the year 2000 to 8–10 billion people in 2050, with 90% of future population growth occurring in developing countries [6]. Over the next three decades, urban growth will bring a further 2 billion people into cities in the developing countries, doubling their size to about 4 billion people. These cities are growing at a rate of 70 million people per year [7].

This growth will result in the creation of mega-cities with populations in excess of 10 million people in each city. In 1950 there was only one mega-city – New York. In 1975 there were 5 and by the year 2015 it is expected that 23 cities around the globe will become mega-cities – 19 of them will be located in developing countries. Table 1.2 shows the 10 largest cities in the world in the year 2000.

These countries already suffer from severe water stress and myriad other social issues. Over 1 billion people or (one in six) live without regular access to safe drinking water. Rapid urbanisation creates squatter towns and slums. For example, currently 40–50% of the population in Jakarta (Indonesia) and a third in Dhaka (Bangladesh), Calcutta (India) and Sao Paulo (Brazil) live in slums [7]. These increases in population will increase the demand for water. Poor sanitation conditions result in increased child mortality. For example, there is one toilet for every 500 people in the slums of Nairobi (Kenya). Leakage rates for most of these cities' water distribution systems are in the high thirties. That is, only two-thirds of the water supplied reaches consumers, whereas in the developed world, approximately 90% of the water supplied reaches the consumer. Poverty also increases the occurrence of water theft – that is, unaccounted for water losses. Figure 1.2 shows the

Table 1.2 **The 10 largest cities in the world in 2000**

City	Country	Population (millions)
Tokyo	Japan	26.44
Mexico City	Mexico	18.07
Sao Paolo	Brazil	17.96
New York	USA	16.73
Mumbai (Bombay)	India	16.09
Los Angeles	USA	13.21
Calcutta	India	13.06
Shanghai	China	12.89
Dhaka	Bangladesh	12.52
Delhi	India	12.44
Buenos Aires	Argentina	12.02
Jakarta	Indonesia	11.02
Osaka	Japan	11.01
Rio de Janeiro	Brazil	10.65
Karachi	Pakistan	10.03

Source: UN Habitat: *Global Urban Observatory*. Nairobi. 2003.

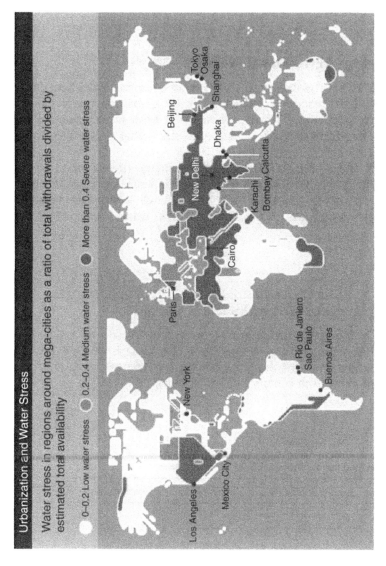

Figure 1.2 The global water challenge – urbanisation and freshwater stress

Source: World Business Council for Sustainable Development – Business in the World of Water. Geneva, Switzerland. August 2006.

urbanisation and water stress in regions around mega-cities as a ratio of total water withdrawals divided by estimated total availability [8].

Whilst almost all the mega-cities are predicted to suffer from water shortages, the problem is particularly acute in China; it is predicted that 550 cities will experience severe water shortages [8].

1.4   Meeting Agricultural Needs

With nearly 70% of global fresh water being used for agriculture (80% in Asia) it will be increasingly difficult to meet global food requirements for a growing population. The development of fresh water resources for human use has compromised natural ecosystems that depend on these resources.

Table 1.3 shows that countries with abundant rainfall, such as in the United Kingdom and France, use relatively small amounts of water for irrigation, whereas countries with low rainfall (which are typically developing countries) use nearly 90% of their water consumption for irrigation.

Case Study: Water Usage in Agriculture in Australia

The water consumption breakdown for year 2000–01 (Figure 1.3) illustrates the heavy use of water in agriculture even though the GDP (gross domestic product) of agriculture has declined significantly from 20% in the first half in the 20th century to 2.9% in 2001–02.

Table 1.3 Sectoral use of fresh water by selected countries

Country	Agricultural (%)	Industrial (%)	Domestic and commercial (%)
India	93	3	4
Egypt	88	5	7
China	87	7	6
Australia*	67	12	12
Netherlands	32	63	5
France	12	71	17
United Kingdom	1	78	21

* Environment flows and water supply accounts for the remainder.
Sources: World Business Council for Sustainable Development Industry. *Fresh Water and Sustainable Development*. April 1998 and Australian Bureau of Statistics *Water Account Australia 2000–01*.

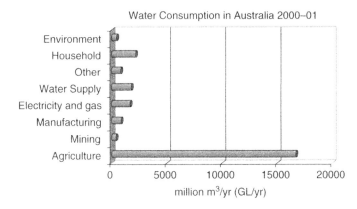

Figure 1.3   Water consumption in Australia 2000–01

Source: Australian Bureau of Statistics.

To cater for an increase in population to 8–10 billion people by 2050, the Food and Agriculture Organisation estimates food demand will double in a similar timeframe. To produce this quantity of food, the additional water use is expected to increase by 50% over the next 30 years to 5600 km^3/yr to eradicate undernutrition and feed an additional 3 billion people. This is almost as much water as the present global consumptive water use in irrigation [9].

Compounding this problem is

- the trend towards water-intensive farming
- over-extraction of water resources
- evaporation of water from open channels
- pollution of these water sources due to heavy reliance on pesticides and fertilizers
- poor land management practices and
- heavily subsidised low water prices encouraging wasteful practices.

Consequently, water has become the number one limiting factor for food production in many parts of Asia and sub-Saharan Africa. The solution is to increase water productivity, that is produce more food from each unit of water. The water used in the production process of an agricultural or industrial product is called the *virtual water* contained in the product [9, 10, 11]. Table 1.4 shows the virtual water requirement per kilogram of some common agricultural products for some selected countries. For example, to produce 1 kg of beef, 13 000 L of water (or more) is required.

Table 1.4 shows that livestock products have a higher virtual water content than cereals and this is understandable. It also shows that USA and Australia are more efficient producers of food than India for the selected products. Export trade in food is in fact *trade in water*. When countries living in water-stressed areas export food, they are in effect exporting water, which further exacerbates the water shortage problem of that country. The largest of the water exporting countries include USA, Canada, Germany and Australia.

Table 1.4 Average virtual water content of some selected products for some of selected countries [9, 10]

Food item	Water requirement m^3/ton		
	USA	Australia	India
Rice (paddy)	1,275	1,022	2,850
Wheat	849	1,588	
Soybeans	1,869	2,106	4,124
Cotton seed	2,535	1,887	8,264
Beef	13,193	16,482	17,112
Pork	3,946	4,397	5,909
Poultry	2,389	7,736	2,914
Lamb	5,977	6,692	6,947

Case Study: Australia's International Trade and Water Exports

Australia's managed water use is 24 000 GL/yr (24 000 million m^3/yr or 19.5 million acre-ft). Australia exports the equivalent of 7500 GL of water/year embodied in goods and services and imports 3500 GL/year [12]. This leaves a net outflow of 4000 GL/year roughly the equivalent to the water consumption of the entire urban sector excluding manufacturing. Given the decreasing contribution that agriculture makes to the Australian economy, the question arises whether the net outflow of 4000 GL/yr is in the nation's long-term interest.

Figure 1.4 shows the virtual water flows in traded crops.

Whilst the problem of under nourishment is a developing-country problem, over capacity and excess of food is a developed-country problem. Obesity, unhealthy food habits, more convenient type foods are driving up water demand in the world. Government subsidies also encourage over-production. In the Organisation for Economic Cooperation and Development (OECD) countries, farmers receive more than one-third of their income from government subsidies, in total over US$300 billion each year [9].

Increases in life styles in the developing countries will further increase meat consumption which means increased water consumption compared to a cereals- or pulses-based diet.

1.5   The Impact of Climate Change

Much has been written about the impact of climate change. Recently there have been a record number of reports providing evidence that climate change is occurring and more importantly the cost of not doing anything to combat it could cost the world trillions of dollars and the extinction of 40% of the

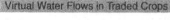

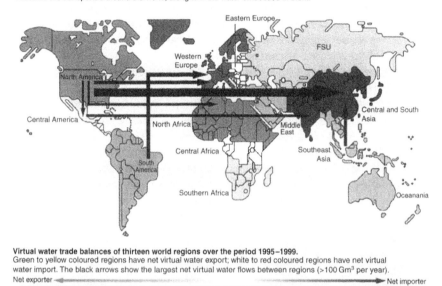

Virtual water trade balances of thirteen world regions over the period 1995–1999.
Green to yellow coloured regions have net virtual water export; white to red coloured regions have net virtual water import. The black arrows show the largest net virtual water flows between regions (>100 Gm³ per year).

Net exporter ◄━━► Net importer

| −1030 | −240 | −140 | −135 | −45 | −22 | −5 | 12 | 20 | 151 | 222 | 380 | 833 |
Net virtual water import, Gm³   1Gm³–10⁶ m³

Source: Adapted form Hoekstra, Hung, and IHE Delft, "Virtual Water Trade," 2002.²⁶

Figure 1.4   Virtual water flows in traded crops

Source: World Business Council for Sustainable Development – Business in the World of Water.
Geneva, Switzerland. August 2006.

species. The principle factors driving climate change are well-documented global warming and greenhouse gas effect.

The question is what impact/effect climate change will have on humans, environment, the economy and the water supplies. A wide-ranging UK study conducted by the former chief economist of the World Bank, Sir Nicholas Stern, is the latest and paints a bleak future if no action is taken [13].

The predicted impacts of global warming include the following:

- Higher maximum temperatures with more hot days and heat waves in nearly all land areas. The earth's surface temperature has increased on average by 0.6° C ever since temperature measurements were started in the 1800s. All of the 10 warmest years have occurred since 1990 including each year since 1995. Climate models indicate a global temperature increase of 1.4–5.8° C (2.5–10.4° F) by 2100 [10]. In Australia the temperatures could increase by 1–6° C [14]. What this means is that a city like Brisbane will have in excess of 20 days with average temperatures above 35° C by 2070 [14].

- Will disrupt traditional rainfall and runoff patterns.
- Up to 40% of species will face extinction.
- Longer and more frequent droughts. Already two regional cities in Australia has almost run out of water. If this pattern continues there is going to be migration of people from these cities to the capital cities which have more resources.
- Extreme weather could reduce global gross domestic product up to 1% [13].
- An increase in severity and more frequent weather-related catastrophes like Hurricane Katrina. For an example, the total economic loss from weather-related catastrophes amounted to US$80 billion out of a total of 216 billion [15]. An increase in ocean temperatures will lead to a doubling of Category 4 and 5 hurricanes. Insurance companies will have trouble covering these disasters due to their severity and frequency.
- Decrease in water quality by changing water temperatures, flows, runoff rates and timing, with significant potential impacts on water users [16].
- Changes in natural water availability will affect water management, allocations, prices and reliability [16].
- Increase in regional conflicts. As the river flows change patterns, regional conflicts amongst countries that share these rivers are going to increase. There are over 2000 regional international treaties sharing water rights in river basins. Some predict that the next world war will be fought over water not oil.
- Impacts on global food production due to global warming, change in rainfall patterns and the increase in carbon dioxide levels resulting in higher food prices. For an example, a more variable monsoon on the Indian subcontinent can impact on the food production of a quarter of a billion people in Bangladesh.
- Hotter, drier summers mean increased demand for water for personal use and air-conditioning. According to a study conducted by the CSIRO, for every degree of global warming, evaporation will increase by 8%.

Case Study: Impacts of Drought in Australia – Grain Harvest Worst in 10 Years

Australia is heading for its smallest harvest in 10 years. Australian Bureau of Agricultural and Resource Economics predicted that in the 2006 financial year economic growth will reduce by 0.7% points. A$6.2 billion would be wiped from the value of farm production, a 35% decrease. Wheat harvest alone has reduced from 20 to 9.5 million tons.

Adapted from: *Sydney Morning Herald*. 30 October 2006. p. 9.

1.6 Business Sector Water Usage

Industry (currently) accounts for 20–22% of the world's consumption of fresh water. As shown in Table 1.3, in industrialised countries this can be as high as 78% and in low-income countries (such as India) 3% of total water use. It is expected that the annual water volume used by industry will rise from $752\,km^3/yr$ in 1995 to an estimated $1170\,km^3/yr$ in 2025 [17].

According to a study conducted by the US investment bank Goldman Sachs by 2050, the economies of the group of countries collectively called BRICs (Brazil, Russia, India and China) could surpass the current G6 countries (United States, Japan, Germany, France, Italy and UK) [18]. If these predictions come true, in US dollar terms, China could overtake Germany by 2007/2008; Japan by 2015 and the United States by 2039. India's economy could be larger than all but the United States and China in 30 years. Russia could overtake Germany, France, Italy and the United Kingdom by 2050.

What impact will these growth rates have on water usage and effluent discharges is not hard to conjure.

Case Study: China

China's GDP growth is a staggering 8%. Its rate of resource consumption particularly water is even greater. For an example, water usage to produce a ton of steel in China is calculated to be 3.8 to 9.3 times more than that of the industrialised countries ($23–56\,m^3$ as against $6\,m^3$/ton of steel in the US and Japan) [8]. Figure 1.5 shows water demand trends and actual wastewater discharges.

Business cannot survive without water and the most important consideration for business is access to clean water. Business uses water in a variety of ways. For example,

- for power and steam generation
- for cooling of air-conditioning systems, process streams and for condensing steam in power plants.
- as a component of product in beverages, pharmaceuticals and so on
- for process needs
- for cleaning vessels and washing floors
- for amenities and
- for irrigation.

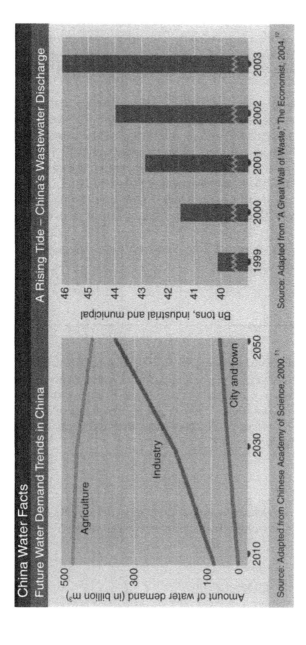

Figure 1.5   China water facts

Source: World Business Council for Sustainable Development – Business on the World of Water, Geneva, Switzerland.

In Sydney, Australia's largest capital city, the business sector consumes approximately 470 million L of water/day. This is approximately 30% of total demand. Of this, industry accounts for 12%, commercial property 10% and government institutions 8% [19].

In contrast to BRICs, in many OECD countries water usage by the industrial sector has been decreasing steadily through the 1980s and 1990s due to economic recession, plant closures, relocation to cheaper sources of labour and a move towards less water-intensive industries.

Not withstanding the efforts to improve industrial water efficiency, it is still at relatively low levels. Renewed efforts are needed to improve water efficiency and reduce water wastage.

Here are a few examples to illustrate the point.

- According to a recent study conducted by the UK Envirowise Agency, British industry and commerce use a staggering 1300 million m³ (1.05 million acre-feet) of water every year – **triple** the amount actually needed for their activities [20].
- Figure 1.6 shows the breakdown in water usage in the food and beverage industry based on 23 water audits carried out by Sydney Water's **Every Drop Counts Business Program**. It is noteworthy that leakage and wastage accounted for 13% – almost equal to the amount of water consumed in the product.
- According to research conducted by the Pacific Institute in California, the commercial and industrial sectors have the potential to save on average 39% and 35% of their water use respectively [21].

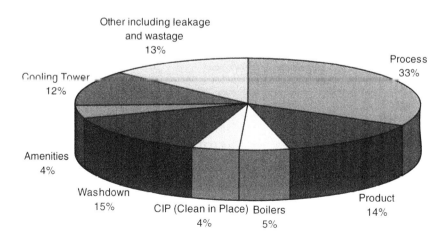

Figure 1.6   Breakdown in water usage within the food and beverage sector in Sydney Courtesy of Sydney Water.

Tables 1.5 and 1.6 shows sectoral usage and savings potential.

• Table 1.7 shows water usage per unit of product among European Union countries [21]. While there might be differences in the mode of data collection, the wide variations in water usage per unit of product support the argument – that even within the developed countries

Table 1.5 **Water usage in the commercial sector in California [21]**

Sector	Current usage (Acre – ft*/yr)	Current usage (ML**/year)	Usage as a percentage (%)	Estimated potential savings (%)
Schools	251,000	309,483	13.5	46
Hotels	30,000	36,990	1.6	33
Restaurants	163,000	200,979	8.8	29
Retail	153,000	188,649	8.3	37
Offices	339,000	417,987	18.3	39
Hospitals	37,000	45,621	2.0	41
Golf courses	229,000	282,357	12.4	36
Laundries	30,000	36,990	1.6	50
Other commercial	621,000	765,693	33.5	38
Total commercial	1,853,000	2,284,749	100	39

*1 Acre ft $= 1233.5\,m^3$
**1 ML (mega litre) $= 1000\,m^3$
Courtesy of Pacific Institute, California, USA.

Table 1.6 **Water usage in the industrial sector in California [21]**

Sector	Current usage (Acre – ft/year)	Current usage (ML/year)	Usage as a percentage (%)	Estimated potential savings (%)
Beverage Processing	57,000	70,281	8.6	16
Dairy processing	17,000	20,961	2.6	29
Fabricated metals	20,000	24,660	3.0	35
Fruit and vegetable processing	70,000	86,310	10.5	26
High tech	75,000	92,475	11.3	39
Meat processing	15,000	18,495	2.3	27
Paper	22,000	27,126	3.3	32
Oil Refining	84,000	103,572	12.6	74
Textiles	29,000	35,757	4.4	38
Other industrial	276,000	340,308	41.5	39
Total Industrial	665,000	819,945	100	35

Table 1.7 Specific water use of industrial production [22]

Country	1 L of beer	1 L of milk	1 kg of cloth	1 kg of paper	1 kg of steel	1 kg of sugar
Austria	10	5	n/d	150	15	15
Denmark	3.4					
France	25	1–4	n/d	250–500	300–600	21–35
Ireland	8				4.5	
Norway	10	1–1.5	130 (all kinds)	20	30	n/d
Spain	6–9	1–5	8–20 (wool)	250	30	3.5–5
Sweden	3–5	1.3	40–50	20	0.6–5.3	0.5
United Kingdom	6.5 (estimated range 2–10)	2.9	6–300 (depends on the type of fabric)	15–30	100	1.5 (estimated range 0.7–6 L)

there are significant opportunities to save water per unit of output. For example, to produce a litre of beer, French beer manufacturers use 25 L of water. Denmark and Sweden use only one-fifth of that amount. Clearly even with an allowance for different production techniques and packaging (draught vs cans), French beer manufacturers could potentially use less water.

1.7   Nine Reasons for Business to Reduce Their Water Consumption

The ultimate objectives of business are to increase profitability and market share. These twin objectives cannot be realised in isolation of the community and the environment. Below are nine reasons why saving water is smart business.

Reason 1   Societal entitlement: Safeguarding security of supply

I am convinced that helping address societal problems is a responsibility of every business, big and small. Financial achievement can and must go hand-in-hand with social and environmental performance
Indra K. Nooyi

Business operates at the local, state and at the global level. The use of water by business is an *entitlement* granted by the community to business. It is a social contract. Especially in times of drought when the community often faces tough water restrictions, business needs to ensure that they too use water responsibly thus honouring the social contract. Consequently, being proactive and using water efficiently will reduce the *water footprint* of any business. When this social contract is not met, governments are likely to take measures to force business to be responsible such as

- imposing limits on production capacity
- more stringent conditions on the use and discharge of the water
- non-renewal of licences and prosecution (e.g. mining sector).

These measures always cost more in the long run – either from

- increased compliance costs for the organisation
- damage to the brand image, or
- loss of public confidence.

Case Study: Mandatory Water Saving Action Plans for High Water Users

Faced with a prolonged drought in Sydney, Australia, the New South Wales State government [23] in 2004 mandated that all businesses using 50 000 m³/yr or more (13.2 million US gal./yr) must develop Water Saving Action Plans and submit them to the regulator of the Department of Energy Utilities and Sustainability. More recently the Queensland government has followed suit with a requirement for business to reduce demand by 25% compared to their 2004/2005 consumption as well develop water-saving plans.

Reason 2   Increasing investor confidence
There is a growing trend to report water usage in company annual reports especially among organisations subscribing to Corporate Social Responsibility (CSR) and Triple Bottom Line (TBL) reporting requirements. CSR is viewed by organisations such as BHP Billion (the largest mining company in the world) as critical to their long-term success. They view their commitment to CSR as reducing their business risks, promotes good business practice.

The UK Environment Agency commissioned Innovest Strategic Value Advisors to research links between sound environmental governance policies and practices, and the financial performance of businesses [24]. The research provides strong evidence of higher financial returns, business opportunity and competitive advantage, with differences in financial performance between environmental leaders and non-leaders as being quite marked. A similar conclusion was arrived at by a Morgan Stanley study. Some examples cited in the Innovest report are given in the case study below.

Inclusion of these companies in indices such as the Dow Jones Sustainability Index (which evaluates companies on their economic performance as well as environmental and social indicators) attracts long-term shareholders – such as pension and superannuation funds and socially responsible mutual funds (Social Responsibility Investment (SRI)). These companies are seen as organisations focussed on long-term financial outperformance by managing and reducing their business, social and environmental risks. The net result is that investors values these companies higher than their peers, thus attracting a higher share price. Australian companies that are rated well include mining giants BHP and Rio Tinto, insurer IAG, brewer Lion Nathan, banks ANZ, NAB and Westpac and building company Lend Lease. A Goldman Sachs study established that well governed companies relative to their peers outperformed by 5–10%.

According to a survey carried out by Calvert (a mutual fund serving 400 000 investors in the USA), 71% of the Americans are more likely to invest in a company that has been rated higher in terms of their social performance and 77% of Americans would purchase more of their products and services [25]. Naturally, an increase in the stock price rewards shareholders and chief executives alike. In the United Kingdom, some estimates suggest that by 2009, 15% of the stock market will be subject to SRI considerations.

Case Study: Winslow Green Growth Fund

The US-based Winslow Green Growth Fund has consistently outperformed its benchmark over a prolonged period with average annual returns above the benchmark index by 20.41, 5.79 and 11.49% over 1, 3 and 5 years respectively.

Case Study: Investa Property Group

Investa Property Group is Australia's largest listed owner of commercial property. Investa has recognised that sustainability practices make good business sense. In 2003, it publicly set environmental performance targets [26] (Table 1.8) and surpassed some of these. In recognition of those efforts in 2004 it was included in the Dow Jones Sustainability World Index.

Table 1.8   Investa environmental targets

Area	Target	Achieved to March 2005
Electricity	15% reduction in usage by June 2006	15.4%
Water	25% reduction in consumption by 2006	26.9%
Waste recovery	50% diverted from landfill Paper/ Cardboard for recycling	Indeterminable 1,019 tons (up from 381 tons)
Aggregate emissions	30,000 tons saved over 3 years to June 2006	22,351 tons
Australian Business Greenhouse Rating Scheme	Average 3.0 stars by June 2005	3.3 stars

Reason 3   Minimising value chain related risk

Today corporate reputation is a more important measure of success than stock market performance. In 2004, The World Economic Forum sent a survey to all 1500 participants at its 34th Annual Meeting that was held in Davos, Switzerland – to understand the issues that concern top business leaders. Out of this, 132 participants represent the world's 1000 leading global companies. Nearly 60% of the survey respondents estimated that corporate reputation represented more than 40% of a company's market capitalisation [27].

For global companies such as Nike and Gap, protecting their brand image means recognising the responsibility to manage not only their own environmental performance but also that of their global suppliers. Therefore they have adopted *global water quality guidelines* for suppliers. This guideline helps protect human health and water quality around the world. As a result, Nike and Gap reap multiple benefits from these programmes.

Reason 4   Cost reduction

Water costs can account for 1–2% of a manufacturing concern's turnover. Thus saving water is an opportunity to reduce costs. A utility's water charges constitute only the *visible costs* of water – tip of the iceberg. In a manufacturing site, water costs are incurred three times. That is,

- at the main meter
- trade waste charges for pollutant strengths and
- sewer usage charges for volumetric discharge.

Additionally, there are other *hidden* costs of water such as

- additional chemical treatment for further purification, boiler, cooling and wastewater treatment
- cleaning costs
- electricity and gas costs
- maintenance costs
- loss of product and
- monitoring and compliance costs.

For example, unrecovered steam condensate wastes energy and chemicals in addition to water. Figure 1.7 illustrates the visible and invisible costs of water.

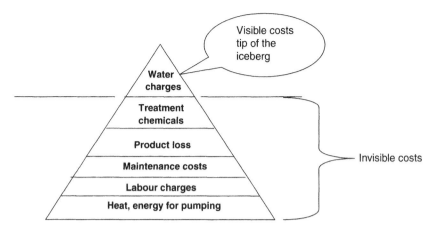

Figure 1.7   Visible and invisible costs of water

Case Study: Unilever Identifies the Hidden Costs of Water

Unilever's Tatura plant in Victoria, Australia, manufactures a range of foods, personal care and household products. Water is used as both an ingredient and a cleaning agent for most production on site. Due to the extremely high quality standards required in food manufacturing, many rigorous processes are in place to ensure standards are met. For example, each time a new product line is produced, all lines must be thoroughly cleaned before production can commence.

To reduce the frequency of cleaning but still ensure line cleanliness, Unilever installed a *pigging* process that negated the use of high-pressure water to clean the lines – hence massive water savings.

The cost savings are

Avoided costs in labour, raw materials and utilities	$360 000
Avoided downtime costs for scheduled cleaning	$200 000
Avoided costs in trade waste charges for BOD	$ 5 000
Total cost savings	**$565 000**
Total cost of implementation	**$1 100 000**
Payback period	**2 years**

Source: *What's New in Food Technology & Manufacturing*. July/August 2004. p. 23–24.

Case Study

A smallgoods manufacturer in Sydney, Australia, estimated that while the water costs are only A\$1.20/m^3, the actual cost of water (once internal chemical treatment, wastewater treatment and sewer discharges were included) was A\$6.00/m^3 – a fivefold increase in costs. These costs are not easily recognisable because they come under different accounts.

In addition to direct production-related costs, there are other compliance costs. For instance, in the food industry, the Hazard Analysis and Critical Control Point system (HACCP) requires certain procedures be followed; similarly, the Environmental Protection Agency (EPA) may require the reduction of certain substances before discharging effluent to the environment. Many of these costs can be minimised by reducing water use.

Reason 5   A counter-measure for future water price increases
Globally, water prices are set to rise over and above inflation. In the United Kingdom in the year 2006/2007, the average price increases 6.5% inclusive of inflation. Historically water prices have been set to recover the capital and operating costs as well as the associated financial costs and dividends to shareholders. Only recently have governments started using the pricing mechanism as a demand management tool to reduce consumption such as having higher tariffs for high water users. Until prices for water reflect its true cost, water wastage will continue. As the available resources becomes scarce, water prices will increasingly reflect the *scarcity value* of water leading to steeper prices [28].

In Sydney, Australia, in 2005, the price of water increased by 20% [29] and a two-tiered pricing system was introduced to residential customers. There was a charge of 20% extra/m^3 – used above normal domestic consumption – to encourage water conservation practices.

Figure 1.8 shows water prices for some selected cities. Not surprisingly the Caribbean has some of the highest prices in the world. Full cost pricing gives a competitive advantage to the more efficient water users over others.

Reason 6   Production efficiency
Using water more efficiently will make additional water available for increased production without necessitating the purchase of additional water, or, in addition – the need to upgrade infrastructure such as pipes, tanks, pumps and other ancillary equipment.

Reporting efficiency benchmark metrics, comparing them with best practice to management and to the media will portray a more positive view of the site, division or company and result in the ability to attract extra resources.

Reason 7   Drives innovation
Water conservation programmes promote innovative thinking within an organisation. It enables the questioning of outdated practices that are redundant but never eliminated. Value chain analysis, trialling of new products and

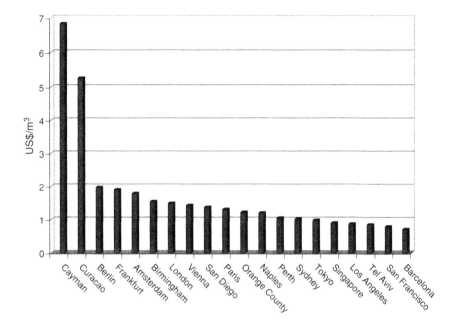

Figure 1.8   The cost of water in major metropolitan areas around the world

Source: International Desalination & Water Reuse Quarterly, November/December 2005, p. 47

processes and adoption of new behaviours drive innovation. For example, by one company selling its wastewater to another, the company can re-value its by-product. What was once a compliance cost has now become a revenue generator. Companies can redesign their products to use less water.

Reason 8   Improved staff awareness and morale
When organisations undertake socially beneficial measures (such as water conservation) there is anecdotal evidence that staff morale improves. Employees are part of the wider community. When the organisation demonstrates socially responsible behaviour, then the employees' values are in alignment with the organisation's values and employees take pride in their employer.

Reason 9   Tax relief and rebates
Many organisations use a hurdle rate of a 2- to 3-year payback as a guide for water conservation–related capital expenditure projects. However, due to the low cost of water, worthwhile water-saving projects often struggle to meet these hurdle rates. Realising this some governments and water utilities provide incentives such as rebates, grants and tax relief to encourage business to undertake water conservation projects that otherwise would have been ignored.

For example, the New South Wales government in Australia recently announced a $120 million Water Saving Fund to assist business to invest in worthwhile water conservation projects.

If businesses can depreciate capital expenditure costs related with these investments at a faster rate than that is currently allowed, then governments do not have to give grants. In Australia, the 1997 Tax Act allows for businesses engaging in agriculture to do so. This scheme if extended to the wider business community will become a catalyst for water conservation–related expenditure with a win-win for all concerned. Such a scheme is in place in the United Kingdom. In the United Kingdom, business can claim investments made towards water conservation as a tax-deductible expense under the Enhanced Capital Allowance scheme (ECA).

1.8   Conclusion

The world's water resources are finite. Increasing demands on water resources by the rapid growth of mega-cities, expansion of agriculture and commercial and industrial growth are depleting available resources. The cost of providing clean water is also increasing as readily available resources are depleted. Climate change – with more frequent and longer droughts – is further straining these meagre resources.

The private sector is good at managing risk. There is no bigger risk than the risk to the environment. Therefore business needs to play a proactive role in reversing this trend and minimising their business risks. It is not about positioning, gaining short-term competitive advantage, feel good media releases and other PR exercises but a genuine assessment of the threat to the business and the environment, the social license, higher water prices, inability to expand production, loss of competitiveness, cost increases, damage to the brand name and loss of community and investor confidence. These considerations will demand a proactive approach to water efficiency be adopted. By doing so, business is doing both its fair share to preserve this finite resource and protecting its market position.

References

[1] Consultative Group on International Agricultural Research. *A Third of the World Population Faces Water Scarcity Today*. Stockholm. www.cgiar.org. 21 August 2006.
[2] World Business Council for Sustainable Development. *Water and Sustainable Development – a Business Perspective*. Switzerland. www.wbcsd.org. 2004.
[3] Biswas A. *An Assessment of Future Global Water Issues*. Third World Centre for Water Management. Water Resources Development. Vol. 21, No. 2. pp. 229–237. June 2005.
[4] World Business Council for Sustainable Development. *Facts and Trends*. www.wbcsd.org. August 2005.

[5] U.S. Water News Online. World's largest aquifer going dry. February 2006.

[6] World Business Council for Sustainable Development. *Industry Fresh Water – Sustainable Development*. www.wbcsd.org. April 1998.

[7] New Internationalist. *Urban Explosion – The Facts*. pp. 18–19. January/February 2006.

[8] World Business Council for Sustainable Development (WBCSD). *Business in the world of water – WBCSD Water Scenarios to 2025*. Conches-Geneva, Switzerland. August 2006.

[9] Stockholm International Water Institute SIWI, IFPRI, IUCN, IWMI. *Let it Reign: The New Water Paradigm* for *Global Food Security*. CSD-13. Stockholm. 2005.

[10] Hoekstra A.Y. and Chapagain A.K. *Water Footprints of Nations*. Vol. 1. Main Report. UNESCO-IHE Delft. The Netherlands. November 2004.

[11] The United Nations Educational. Scientific and Cultural Organisation. *Water a Shared Responsibility. The United Nations World Water Development Report 2*. Berghahn Books Inc. New York. 2006.

[12] Foran B. and Poldy F. Chapter 6 – *The Future of Water*. CSIRO Sustainable Ecosystems. Canberra. October 2002.

[13] BBC News. At-a-glance: The Stern Review. www.bbc.co.uk 30 October 2006.

[14] Department of Parliamentary Services. *Issues Encountered in Advancing Australia's Water Recycling Schemes*. Parliament of Australia. Canberra. 16 August 2005.

[15] Mills E. and Lecomte E. From Risk to Opportunity: *How Insurers Can Proactively and Profitably Manage Climate Change*. CERES. Boston MA. August 2006.

[16] Morison J. and Gleick P. *Freshwater Resources: Managing the Risks Facing the Private Sector*. Pacific Institute. Oakland, California. August 2004.

[17] UN World Water Development Report. *Water for People – Water for Life*. 2001.

[18] Goldman Sachs *Dreaming With BRICs: The Path to 2050*. Global Economics Paper 99. Global Economics Weekly. www.gs.com.

[19] Sydney Water. *Water Conservation & Recycling – Implementation Report 2004–2005*. Sydney 2005.

[20] Envirowise Media release. *UKs Desert State – Envirowise urge businesses to curb water usage: UK Businesses are wasting three times too much water*. www.envirowise.gov.uk. August 2005.

[21] Gleick P. *Waste Not Want Not – The Potential for Urban Water Conservation in California*. Pacific Institute. Oakland, California. November 2003.

[22] European Environment Agency. *Sustainable Water Use in Europe: Part 1 – Sectoral Use of Water*. Brussels. 1999.

[23] Department of Natural Resources and Planning. *NSW Govt Securing Sydney's Water Future*. Sydney. 2004.

[24] The Environment Agency. *It Pays to be Green. Environment Agency tells Business and Investors*. UK. 9 November 2004.

[25] Calvert Survey. *55% of Investors Think SRI Mutual Funds Help Keep Companies Honest, Products Safer*. Yahoo Finance. Press Release. 23 January 2006.

[26] Investa Property Group. *2005 Financial Report*. Sydney. 2005.

[27] World Economic Forum Press Release. *Corporate Brand Reputation Outranks Financial Performance as the Most Important Measure of Success*. 22 January 2004.

[28] Tom J. *Pricing Water*. OECD Environment Directorate. March 2003.

[29] Independent Pricing Tribunal. *Pricing Determination*. 2005.

[30] Holliday C.O., Schmidheiny S. and Watts P. *Walking The Talk – The Business Case for Sustainable Development*. Greenleaf Publishing. 2002.

Chapter 2

Basic Water Chemistry

2.1 Overview

Before we delve into water conservation, it is useful to have an understanding of the basic concepts in water chemistry and ions commonly found in water.

Water is known as the *universal solvent* due to its power to dissolve virtually all substances to some extent. The key to this phenomenal ability is its *structure*. The water molecule consists of two atoms of hydrogen and one atom of oxygen with a slight positive charge near the hydrogen atoms and a slight negative charge near the oxygen atom. This polarity enables the water to dissolve all ionic and polar substances. Non-polar compounds such as hydrocarbons are insoluble in water.

For this reason *pure water* is not found in nature. It always contains impurities, which impart colour, clarity, taste, smell and feel. The types of impurities found in water and wastewater can be divided into four groups: dissolved, physical, microbial and radiological (Table 2.1).

Impurities in natural waters depend on the source of water. Wells and spring waters are classed as *ground water* while rivers and lakes are known as *surface waters*.

Ground water picks up impurities as it seeps through the rock strata, dissolving some parts of (or almost all) it makes contacts with. However, the natural filtering effect of rock and sand usually keeps water free of suspended matter.

Wastewater may contain a myriad of substances depending on the source and can range from inorganic to organic substances. For example, effluent from food processes have high levels of organic matter, and leachate from landfills contains organic substances, ammonia as well as heavy metals.

Solubility principles, some common substances found in commercial and industrial water systems, and water-quality guidelines are discussed below. It is by no means a complete list; however, it gives the reader an appreciation for the types of contaminants and their impacts on water systems.

Table 2.1 Types of impurities found in Water

Types of impurities	Examples
Dissolved	
Positive ions (cations)	Calcium, magnesium, sodium
Negative ions (anions)	Carbonate, bicarbonate, phosphate, chloride, cyanide, nitrate
Disinfectants	Chlorine
Heavy metals	Copper, nickel, chromium, lead, mercury
Gases	Oxygen, carbon dioxide, ammonia, hydrogen sulphide
Physical	
Colour	Dissolved organic matter, iron, manganese, dyes, algae
Taste and odour	Geosmin, sulphide, chlorine
Appearance	Silt-suspended solids, plankton, oil fats, petroleum hydrocarbons
Microbial	
Bacteria	*E. Coli, Campylobacter, Legionella*
Viruses	Adenovirus, Reovirus, Hepatitis A
Protozoa	*Cryptosporidium, Giardia*
Other	Cyanobacteria (blue-green algae)
Radiological	
Naturally occurring	Radium, uranium

2.2   Solubility Principles

The solubility of a substance in water is a function of

- temperature
- pressure
- pH
- redox potential and
- the relative concentrations of other substances in solution.

In a natural environment these variables are related in such a complex manner that exact solubilities cannot always be predicted. However, some general rules can be formulated.

Common dissolved substances can be segregated into gases and minerals. Dissolved gases are primarily oxygen and carbon dioxide. Sometimes ammonia will be present in certain wastewaters. Ground water may contain hydrogen sulphide gas. Their concentrations are typically expressed in milligrams/litre (mg/L) or in parts per million (ppm).

2.3   Common Substances Found in Water

2.3.1   pH

Water molecules ionise into H^+ and OH^-. The concentrations of these ions are very small. The concentration at room temperature is 1×10^{-7} grams of per litre or one part of hydrogen ion to 10 million parts of water. This concentration can also be expressed as

$$pH = -Log\ [H+] = -Log\ (1 \times 10^{-7}) = 7 \tag{1}$$

The pH of 7 is considered neutral. A pH between 0.0 and 7 is acidic. A pH between 7 and 14 is basic. Strong acids have a pH of 1, which means one part H^+ ion in 10 parts of water. A change in pH to 2 is a tenfold decrease in H^+ concentration, which equates to one part H^+ in every 100 parts of water. This simple concept is important in understanding the changes ionic substances have on water.

The majority of living organisms operate at a pH of 7 and natural water has a pH between 6.5 and 9.5. Drinking-water guidelines generally specify a pH of 6.5–8.5. Seawater has a pH close to 8 and acidic hot spring water has a pH of 1.8. The normal pH for irrigation water is 6.5–8.4. Dissolved substances, temperature, microbial activity, however, can change this balance. Figure 2.1 shows these pH ranges for the different types of water.

pH has a strong influence on

- corrosion of metals
- concrete and other infrastructure
- scaling potential inside pipes
- heat exchangers or boiler tubes and
- the mobility of metal ions.

Generally pH < 6 is corrosive to metal piping and pH > 8 may cause scaling (due to precipitation of calcium and other metal ions with anions such as carbonates, phosphates and sulphates). These effects can be magnified in the presence of dissolved gases and ions. For this reason, many industrial water systems are generally controlled between a pH range of 7 and 10 including trade waste discharges to the sewer.

Only a few ions such as sodium (Na^+), potassium (K^+), nitrate (NO_3^-) and chloride (Cl^-) remain in solution through the entire range of pH values found in water.

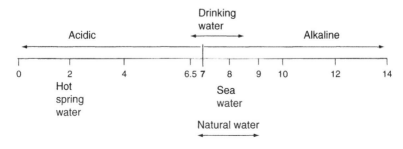

Figure 2.1   pH scale

2.3.2   Dissolved Gases

The dissolved gases of interest to the water chemist are carbon dioxide, oxygen, ammonia and hydrogen sulphide. These gases are discussed below.

2.3.2.1   Carbon dioxide and Alkalinity

The atmosphere contains 78% Nitrogen (N_2), 21% Oxygen (O_2) and 0.033% Carbon dioxide (CO_2).

Despite the low concentration of CO_2 in the atmosphere, due to its higher solubility (about 30 times that of oxygen), it plays a major role in water chemistry.

Alkalinity is a measure of the buffering capacity of water or the capacity of bases such as carbonates to neutralise acids.

CO_2 dissolves in water to form carbonic acid (H_2CO_3) decreasing the pH and increasing the acidity of the water. As such, at a pH of 4.2–4.5 and below, all CO_2 is in the form of H_2CO_3 and no alkalinity is present. Above pH of 4.5, H_2CO_3 is present in the form of bicarbonates (HCO_3) and carbonates (CO_3). These reactions are shown below.

$$CO_2 + H_2O \longleftrightarrow H_2CO_3 \longleftrightarrow H^+ + HCO_3^-$$

At a pH greater than 4.5, alkalinity is a combination of H_2CO_3 and HCO_3^-. As the pH increases, the HCO_3^- ion becomes the dominant ion.

At pH greater than 8.2–8.4 there is no more free CO_2 gas left – and all alkalinity is in the form of HCO_3^- ion and the more alkaline CO_3^{2-} ion.

$$HCO_3^- \longleftrightarrow CO_3^{2-} + H^+$$

At pH greater than 9.6, no HCO_3^- is present and all alkalinity is in the form of the CO_3^{2-} ion. Further increases in pH results in hydroxyl ions also appearing and therefore hydroxyl alkalinity can be measured above pH of 9.6.

The total alkalinity of water $= \sum HCO_3^- + CO_3^{2-} + OH^-$ concentrations.

Table 2.2 and Figure 2.2 shows how the pH of a solution dictates the form of alkalinity that will be dominant in the solution.

This remarkable ability of CO_2 to exist in many forms gives the water its unique buffering capacity. In a buffer solution, the pH varies only marginally with the addition or removal of H^+. Waters with low alkalinity are very

Table 2.2   Forms of alkalinity

pH	H_2CO_3/CO_2 gas	HCO_3^-	CO_3^{2-}
<4.2–4.5	dominant	absent	absent
4.3–8.3	present	dominant	absent
>8.3	absent	present	dominant
>9.6	absent	absent	dominant

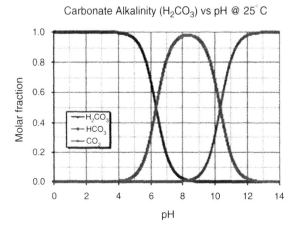

Carbonate Alkalinity (H_2CO_3) vs pH @ 25° C

Figure 2.2 Alkalinity and pH relationship

susceptible to changes in pH. Waters with high alkalinity are able to resist major shifts in pH. High alkalinity is also desirable in boiler water to minimise steel corrosion and scaling of some ions.

Measurement of alkalinity is done by titration. Titrations for alkalinity can also include other ions such as borates, phosphates and silicates and other organic substances. It is customary to express alkalinity as mg/L $CaCO_3$.

2.3.2.2 Oxygen (O_2)
While oxygen is required for living organisms, it is the principal corrosive in water systems. It causes pitting corrosion, a form of localised corrosion which leads to rapid failures of pipes and heat exchangers. Solubility of oxygen in water is a function of temperature and pressure. This property is used to remove oxygen in steam systems.

Refer to Chapters 5 and 7 for more details.

2.3.2.3 Ammonia (NH_3)
At high pH, ammonia is present as NH_3 gas. At low and neutral pH, ammonia is found as the ammonium ion (NH_4). Natural levels of ammonia in surface water and ground water are usually below 0.2 mg/L [1]. Above 1.0 mg/L, ammonia becomes objectionable in drinking water. In raw sewage, ammonia concentrations are in excess of 10 mg/l. For this reason, the presence of ammonia in water is a possible indication of bacterial, sewage and animal waste pollution. Ammonia in industrial water systems is present as the ammonium ion, or as organic nitrogen in industrial effluent such as landfill leachate, petrochemical plants, food processing plants and in steam condensate systems. It is corrosive to copper and its alloys. NH_3 in the presence of oxygen contributes to accelerated corrosion of copper alloys – even as little as 0.2 mg/L can cause corrosion of admiralty (70/30 Cu/Zn) heat exchangers.

2.3.2.4   Hydrogen Sulphide (H_2S)

The presence of hydrogen sulphide gas is detected by its *rotten egg* smell with a threshold odour level of 0.005–0.025 mg/L [2]. At concentrations of 3–5 mg/L it is considered to be offensive. It is a highly toxic gas which overwhelms the sense of smell (loss of smell occurs at 100–150 mg/L) and go onto overcome the victim and cause death. Found in some well, marshy and wastewaters it is readily oxidised to the elemental sulphur.

2.3.3   Dissolved Ions

Fresh water always contains dissolved ions, which come from the dissolution of minerals such as limestone, magnesite, gypsum and decaying plant materials. The most common ions found in water are given in Table 2.3.

To a lesser extent other cations are present such as

- Iron (Fe^{2+} and Fe^{3+})
- Manganese (Mn^{2+})
- Aluminium (Al^{3+})
- Ammonium (NH_4^+) and
- Copper (Cu^{2+}).

Less common anions are

- Carbonates (CO_3^{2-})
- Hydroxide (OH^-)
- Sulphides (S^{2-}) and
- Phosphates (PO_4^{3-})
- Silicate (SiO_4^{2-}).

2.3.3.1   Conductivity and Total Dissolved Solids

Conductivity is a measurement of the ability of water to conduct electricity due to salts present in water. The higher the dissolved salts content of the water, the higher its conductivity. Monovalent salts have a higher conductivity than divalent salts. Organic compounds like oil, phenol, alcohol and sugar are not ionized in water and therefore have a low conductivity. Conductivity is measured in micro Siemens per centimetre ($\mu S/cm$). Distilled water has conductivity in the range of 0.5–3 $\mu S/cm$. Good-quality drinking water has a conductivity less than 700–800 $\mu S/cm$, [1, 3] whilst animals such as sheep can tolerate

Table 2.3   Common ions found in water

Cations (positively charged ions)	Anions (negatively charged ions)
Calcium (Ca^{2+})	Chlorides (Cl^-)
Magnesium (Mg^{2+})	Bicarbonates (HCO_3^-)
Sodium (Na^+)	Sulphate (SO_4^{2-})
Potassium (K^+)	Nitrates (NO_3^-)

Table 2.4 Comparison of drinking water and mine water to Australian Drinking Water Guidelines (NHMRC) maximum limits

Test description	NHMRC Guidelines	Drinking water Sydney	Mine water
pH	6.5–8.5	7.5–8.9	7.4
Total dissolved salts	500	61–130	8100
Colour (True)	15	<3	
Turbidity	5	0.1–0.5	
Total hardness as CaCO3	200	40–68	917
Aluminium as Al	0.2	<0.05	
Ammonia as NH3	0.5	0.01–0.45	0
Arsenic as As	0.007	nd*	0.054
Cadmium as Cd	0.002	nd*	0.001
Chloride as Cl	250	15–90	1400
Chromium (VI) as Cr	0.05	nd*	0.3
Copper as Cu	1	0.002–0.25	0.003
Cyanide as CN	0.08	nd*	nd
Fluoride	1.5	0.9–1.2	
Iron as Fe	0.3	0.01–0.09	10
Lead as Pb	0.01	nd*	0.004
Magnesium as Mg	na**	1.5–6.2	120
Manganese as Mn	0.1	0.001–0.005	0.35
Mercury as Hg	0.001	nd*	<0.001
Nitrate as NO_3	50	0.04–5.6	10
Nitrite as NO_2	3	<0.003–0.5	nd
Selenium as Se	0.01	nd*	0.017
Silver as Ag	0.1	nd*	nd
Sodium as Na	180	4–53	1600
Sulfate as SO_4	250	1–20	430
Zinc as Zn	3	<0.008	0.007
Organic compounds	Various	nd*	1500
Free chlorine	5	0.1–1.2	nd
Disinfection by products (trihalomethane)	0.25	0.02–0.11	nd

NB: All units expressed as mg/L or otherwise stated. pH, colour and turbidity are expressed as units, Hazen units and as nephelometric units respectively.
* nd – none detected
** nd – no health or aesthetic value in guideline

higher conductivities up to 16 500 μS/cm. Table 2.4 shows the National Health & Medical Research Council's Australian Drinking Water Guidelines (NHMRC) compared to drinking water quality in Sydney and mine water.

Certain well waters and brackish water (intrusion of sea water into low conductivity water) has a conductivity of 10 000–23 000 μS/cm. Seawater has a conductivity of 50 000 μS/cm or more.

Conductivity of industrial effluent such as landfill leachate can be as high as 14 000 μS/cm. On the other hand, water suitable for use in the electronics and microelectronics industry for purposes such as the washing and rinsing of semiconductor components in cleaning and etching operations require extremely high purity water and for this reason is expressed as resistivity megaohm-cm. It is the inverse of conductivity. A Ultrapure water (type E-1) has a resistivity of 18 megaohm-cm. The dissolved contaminants are measured in μg/L or ng/L (parts per billion or trillion). Thermal power plants too require high-quality water in their high-pressure steam systems (Chapter 7 discusses steam system water-quality requirements).

The relationship of conductivity and total dissolved solids (TDS) is given below:

TDS $= 0.55 - 0.8 \times$ conductivity. The average is 0.64.

Other units of expressing TDS are given in Table 2.5.

The dissolved solids content of water can adversely impact plants ability to absorb water from the soil. Most plants cannot tolerate conductivities in excess of 2300 μS/cm. Table 2.6 shows guidelines for salinity of irrigation water [4].

Some common salts and ions that contribute to dissolved solids are discussed below.

2.3.3.2 Hardness, Calcium and Magnesium

Calcium (Ca^{2+}) and magnesium (Mg^{2+}) give water its *hardness*. Hard water is difficult to lather and therefore *hardness* is a measure of Ca^{2+} and Mg^{2+} ions. Calcium typically accounts for two-thirds of total hardness in surface water and is usually in the range of 2–200 mg/L as Ca. Sea water may contain as much as 400 mg/L as Ca [1]. The typical concentration for Mg^{2+} in surface water is about 10–50 mg/L as Mg. However, in well and in sea water Mg^{2+} can be about four to five times that of Ca^{2+}. Hardness is typically expressed as mg/L $CaCO_3$. Most common Ca^{2+} and Mg^{2+} salts are bicarbonates, sulphates, chlorides and nitrates. Other alkaline earth elements of this group of interest are barium and strontium. Typically barium is found in some well waters with typical concentrations between 0.05 and 0.2 mg/L. Strontium is found in certain well waters. Typical concentrations are less than 15 mg/L.

Salts of Ca^{2+} and Mg^{2+} (except for Cl^-, NO_3^-, HCO_3^-) are sparingly soluble or insoluble in water. The solubility product (K_{sp}) of a substance shows the solubility of a given substance in water and is a constant. The higher the K_{sp} of a substance the greater its solubility in water. The K_{sp} for

Table 2.5 Common conversion factors for expressing total dissolved solids

Unit	To convert	Multiply by	To obtain
Conductivity	μS/cm (microS/cm) or μmho/cm		
	(when <5000 μS/cm)	0.64	mg/L
	(when >15000 μS/cm)	0.8	mg/L
TDS	mg/L	1	ppm
Water salinity	deciSiemens/metre (dS/m) or millimho/cm	1000	μS/cm or μmho/cm

Table 2.6 Guidelines for salinity of irrigation water

Class	Comment	Electrical conductivity μS/cm	Total dissolved solids mg/L
1	*Low-salinity.* Water can be used with most crops on most soils with little likelihood that a salinity problem will develop.	0–280	0–175
2	*Medium-salinity.* Water can be used if moderate leaching occurs. Plants with medium salt tolerance can be grown, usually without special control for salinity control. Special care needs to be taken when irrigating salt-sensitive crops with sprinkler.	280–800	175–500
3	*High salinity.* Water that cannot be used on soils with restricted drainage. Special considerations required even with adequate drainage.	800–2,300	500–1,500
4	*Very high salinity.* Water not suitable for irrigation water under ordinary conditions. Soil must be permeable and salt-tolerant crops must be selected.	2,300–5,500	1,500–3,500
5	*Extremely high salinity.* Water may be used only on permeable, well-drained soils with good management practices.	>5,500	>3,500

Adapted from Agriculture and Resource Management Council of Australia and New Zealand. *Guidelines for Sewerage Systems – Reclaimed Water.* February 2000.

$CaCO_3$ is 4.8×10^{-9}, indicating that it is not very soluble in water when compared to a K_{sp} 36 for common salt (NaCl). When the concentration of $CaCO_3$ exceeds its solubility product, K_{sp} limit, it precipitates. This is shown by

$$[Ca^{2+}] \times [CO^{2-}_3] > K_{sp}[CaCO_3] > 4.8 \times 10^{-9}$$

As Table 2.7 shows there are less soluble substances than $CaCO_3$

The Ca^{2+} and Mg^{2+} ions react with various anions such as SO_4^{2-}, CO_3^{2-} and PO_4^{3-} to form scale in various applications such as on reverse osmosis membranes, boiler tubes and in cooling water heat exchangers. The greater the hardness of water, the greater the scaling potential of the water as shown in Table 2.8.

Build up of scale reduces heat transfer in heat exchangers, boiler tubes and other heat transfer equipment leading to production losses and more frequent turnarounds. For example, in a water tube boiler, a scale thickness of 0.8 mm (1/32 in.) can result in heat loss of 8% and increased fuel consumption of 2%.

Another characteristic of these salts is that unlike most water-soluble salts they have an inverse relationship with temperature. Salts such as sodium

Table 2.7 Solubility products for some common scalants at 20° C

Salt	Formula	Solubility product, K_s
Calcium sulphate	$CaSO_4$	2.3×10^{-4}
Magnesium carbonate	$Mg\,CO_3$	1.0×10^{-5}
Calcium hydroxide	$Ca(OH)_2$	8.0×10^{-6}
Calcium hydrogen phosphate	$CaHPO_4$	2.0×10^{-7}
Calcium carbonate	$CaCO_3$	4.8×10^{-9}
Barium sulphate	$BaSO_4$	9.2×10^{-11}
Calcium fluoride	CaF_2	3.2×10^{-11}
Magnesium hydroxide	$Mg(OH)_2$	3.4×10^{-11}
Manganese hydroxide	$Mn(OH)_2$	4.0×10^{-14}
Manganese sulphide	MnS	1.4×10^{-15}
Ferrous hydroxide	$Fe(OH)_2$	4.8×10^{-16}
Ferrous sulphide	FeS	4.0×10^{-19}
Aluminium hydroxide	$Al(OH_3)$	8.5×10^{-23}
Ferric hydroxide	$Fe(OH)_3$	3.8×10^{-38}

Adapted from Judd S. and Jefferson B. *Membranes for Industrial Recovery and Reuse.* 2003 [5].

Table 2.8 Relationship between hardness and scaling potential of water [3]

Hardness as mg/L $CaCO_3$	Description
<60	Soft but possibly corrosive
60–200	Good quality
200–500	Increasing scaling problems
>500	Severe scaling

chloride (NaCl) increase in solubility with temperature. $CaCO_3$ solubility, however, decreases with a rise in temperature. This decreasing solubility with an increase in temperature is the reason for scale formation on the hottest surfaces of heat exchangers or boiler tubes.

In cooling water or boiler water analyses, expressing ions as $CaCO_3$ is particularly useful.

To convert an ion into the $CaCO_3$ form:

$$\text{Ion as } CaCO_3\,mg/L = \frac{\text{Concentration of } M^n \text{ mg/L} \times 50}{\text{Equivalent weight of } M^n}$$

where

50 – Equivalent weight of $CaCO_3$
M^{n+} – represents any ion (cation or anion).

Example 1

Calculate the total hardness in mg/L $CaCO_3$ for a solution with the following cation concentrations shown in Table 2.9

$$Ca^{2+} \ (CaCO_3 mg/L) = 80 \times 50/20 = 200$$

$$Mg^{2+} \ (CaCO_3 mg/L) = 35 \times 50/12.2 = 143.4$$

The total hardness of the solution $= 200 + 143.3 = 343.4$ mg/L $CaCO_3$.

Another useful conversion is that from mg/L to milliequivalent per litre (meq/L). This is derived at by dividing the concentration by the equivalent weight of the ion or substance.

$$Milliequivalent/L = \frac{Concentration \ of \ ion \ in \ mg/L}{Equivalent \ weight \ of \ ion \ or \ substance}$$

where the equivalent weight = atomic weight/valency

Worked example

Convert 100 mg/L of Ca^{2+} to meq/L. The atomic weight is 40 and valency is 2.

To convert 100 mg/L to milliequivalent/L $= 100/20 = 20$ meq/L.

2.3.3.3 Chlorides

Chloride (Cl^-) is highly soluble and for this reason is a common constituent of water. In drinking water, Cl^- becomes noticeable above 200 mg/L [1] and the World Health Organisation recommended upper limit in drinking water is 250 mg/L. Cl^- ions accelerate corrosion of metals especially stainless steel even at concentrations as low as 50 mg/L. To mitigate against the possibility of stress corrosion cracking of austenitic stainless steels, limits are put on Cl^- concentrations. For type 304 stainless steel at temperatures below 60° C, the Cl^- concentrations are limited to 200 mg/L [6] and for type 316–1000 mg/L. Generally the more susceptible a metal is to general corrosion; the less susceptible it is to Cl^- attack [7].

Cl^- also impact on salt-sensitive plants especially with sprinkler irrigation. Generally a Cl^- concentration less than 70 mg/L is safe for all plants and concentrations above 350 mg/L can cause severe problems [8].

Table 2.9 Cation concentrations

Cation	Concentration mg/L	Molecular weight (MW)	Valency (V)	Equivalent weight = MW/V
Ca^{2+}	80	40	2	20
Mg^{2+}	35	24.3	2	12.2

A related ion is fluoride ions. Normally injected to municipal water supplies at 2.5 mg/L, fluoride levels in excess of 5 mg/L can cause mottling of teeth.

2.3.3.4 Sodium
Sodium (Na$^+$) salts are highly soluble and do not cause scaling but cause corrosion of metals and at high concentrations reduce the clay-bearing soil's permeability as well as affecting the soil structure. Sodium has the capacity to disperse clay particles thus making the soil more prone to crusting and reducing its permeability. The sodium adsorption ratio (SAR) shows the degree of sodium adsorption by a soil from a given water and is expressed as:

$$SAR = \frac{Na^+ \; meq/L}{\sqrt{[(Ca^{2+} \; meq/L + Mg^{2+} \; meq/L)/2]}}$$

Water with SAR values greater than 18 is unsuitable for continuous irrigation [8].

2.3.3.5 Iron
Although less common than Ca^{2+} and Mg^{2+}, Iron (Fe) occurs naturally in water, usually at <1 mg/L, but can be as high as 100 mg/L [3] in oxygen-depleted well water especially below a pH of 7. Above 0.3 mg/L iron imparts a metallic taste to water and for this reason drinking water is limited to a maximum of 0.3 mg/L.

High concentrations of iron

- stains laundry and fittings in toilets
- forms deposits and corrosion in pipes, boilers and cooling systems and
- causes odours.

Iron exists in two forms. The water-soluble form is known as ferrous (Fe^{2+}). In non-aerated well waters, Fe^{2+} behaves much like Ca^{2+} and Mg^{2+} in that it can be removed by ion exchange or through precipitation. Upon aeration the Fe^{2+} gets converted to the more insoluble ferric (Fe^{3+}) form. Being colloidal in nature, Fe^{3+} in concentrations as little as 0.05 mg/L iron can cause fouling in paper mills, ion exchange resins and front end of reverse osmosis (RO) membrane systems.

2.3.3.6 Manganese
Manganese (Mn) is a water contaminant present in both well and surface waters with levels up to 3 mg/L. Like iron Mn also imparts a metallic taste to water. It is present in organic complexes in surface waters. Like Fe it is soluble in oxygen-free water. Upon exposure to air forms black deposits (MnO$_2$). It creates stains with concentrations in excess of 0.1 mg/L [3]. The limit for drinking water is 0.05 mg/L. Similarly in RO membrane systems it is controlled below 0.05 mg/L in the feed water.

Manganese also causes pitting corrosion of stainless steel. In beverage and food processing industry, Mn can impair taste and promote chemical deterioration of products [9].

2.3.3.7   Silica

Silicon is the second most abundant element on the surface of the Earth. Most common is silica (SiO_2), ranging from 1 to 100 mg/L in natural waters. Higher concentrations are found in waters with significant volcanic activity. The *total silica* levels are comprised of *reactive silica* and *unreactive silica*.

Reactive silica (silicates SiO_4) is dissolved silica that is slightly ionised and has not been polymerised into a long chain. Silicate scale is formed when soluble silica reacts with Ca^{2+} and Mg^{2+} to form various complex silicates at high temperatures. For this reason silica concentrations are controlled below 150 mg/L in cooling water systems. In boiler water, silica concentrations are a function of pressure.

Unreactive silica is polymerised (or colloidal silica), acts more like a solid than a dissolved ion. In steam systems, silica as SiO_2 forms glassy deposits on steam turbine blades. In RO membrane plants colloidal silica causes fouling of the front end of the membrane elements.

2.3.3.8   Phosphate

Phosphate (PO_4^{3-}) is found in domestic and industrial wastewaters. PO_4 contributes to algal blooms and for this reason most modern laundry formulations do not contain phosphate. In industry, PO_4 are used in cooling water, boiler water treatment and in the metal finishing industry. In high concentrations, PO_4 forms calcium phosphate scale.

2.3.3.9   Nitrate

Nitrate (NO^-_3) ions are highly soluble and are present as a result of the nitrogen cycle. Found naturally, high concentrations are an indication of biological decomposition of organic waste or due to agricultural runoff. The recommended drinking water limit for NO^-_3 is 44–100 mg/L as NO^-_3. To express NO^-_3 as nitrogen (N), divide the NO^-_3 concentration by 4.43.

2.3.3.10   Boron

Boron in freshwater is normally present at concentrations of 0.1 mg/L and in sea water it is found at concentrations of 5 mg/L. In recycled water (from laundry detergents containing perborate formulations) and industrial run off, boron is present at much higher concentrations. It is essential to plants at low concentrations but toxic at high concentrations as much as 0.75 mg/L for onions and peach [8].

2.3.3.11   Cyanide

Cyanide (CN^-) is found in many industrial wastewater streams such as in plating, metal cleaning, is a by product of coke oven gas processes and used in gold-mining applications. As the acid HCN it is highly toxic.

2.3.4   Suspended Solids and Turbidity

Suspended solids are referred to as suspended and colloidal sediment matter or coarse particles. These solids settle out of standing water. The concentration of suspended solids in water is determined by passing a known quantity of water through a 0.45-μ membrane filter and determining the weight of solids collected on the filter. The weight of solids collected divided by the quantity of water passed through the filter gives the suspended solids concentration in mg/L.

Suspended solids contribute to the fouling of process and heat transfer equipment. Suspended solids can harbour micro-organisms which can result in health-related risks as well as corrosion of materials known as microbiologically influenced corrosion.

Turbidity is a measure of the cloudiness of water. It is an indirect measurement of suspended solids. The cloudy appearance is caused by the scattering and absorption of light by these particles. Turbidity cannot be directly correlated with suspended solids concentrations since the light scattering properties of suspended matter vary, and turbidity measurements only show the relative resistance to light scattering not an absolute measurement of concentration of suspended matter. Low turbidity concentrations are commonly expressed as Nephelometric Turbidity Units (NTU). It measures the intensity of light scattered at 90° as a beam of light passes through a water sample. Clear water has a value of 1 NTU. Slightly muddy water has NTU of 5. The drinking water standard for turbidity is below 5 NTU. Turbidity in excess of 1 NTU may shield some micro-organisms from disinfection. Steam systems also require water with very low turbidity.

Other units of turbidity are Jackson Turbidity Units (JTU) and Formazin Turbidity Units (FTU). They are not correlated.

2.3.5   Colour

The colour of natural water is influenced by decaying organic matter such as tannins, lignin, iron or manganese salts. Dye house or paper mill effluent frequently gives rise to coloured water. Colour due to organic substances such as fulvic and humic acid poison ion exchange resins in steam systems and causes brightness problems in the paper industry. Colour is typically measured as Hazen units (HU) or in Pt/Co units. 1 Pt/Co unit is equal to 1 Hazen unit. Drinking water standards have adopted the Hazen unit and 10 Hazen units are barely noticeable in a glass of water. The limit for drinking water is less than 15 HU [1, 3, 10].

Natural waters range from <5 in very clear waters to 1200 mg/L Pt in dark, peaty waters [10]. For true colour measurement the water samples need to be filtered using a 0.45-μ filter to eliminate the light-scattering effect of suspended solids.

2.3.6   Organics in Water

Organics are found in freshwater and wastewater systems. These can be categorised as soluble organics, which include natural and synthetic organic chemicals. Most naturally occurring organic matter are negatively charged colloids and will be degraded by the action of micro-organisms. Some man-made organics do not degrade naturally; these are known as bio-refractory compounds and for this reason regulatory agencies strictly limit their discharge. Examples are aromatic and chlorinated hydrocarbons and dioxins from paper mill effluent.

Organic concentrations are typically expressed as biochemical oxygen demand (BOD) or chemical oxygen demand (COD).

2.3.6.1   Biochemical Oxygen Demand
Biochemical oxygen demand (BOD) is the amount of oxygen in mg/L used by micro-organisms to consume biodegradable organics in wastewater under aerobic conditions over a 5-day period at a temperature of 20° C. The 5-day test equates to 2/3 of the total BOD demand. Regulatory authorities require that BOD be below 10 mg/L for most reuse applications.

2.3.6.2   Chemical Oxygen Demand
Chemical oxygen demand (COD) measures the total organic content that can be oxidised by potassium dichromate ($K_2Cr_2O_7$) in a sulphuric acid solution. Some organic substances such as aromatic hydrocarbons and pyridines are not oxidised by the COD test under any circumstances. Nitrogen compounds are also not oxidisable. Therefore this test may give falsely low readings where these substances are present. The attractiveness of COD over BOD is that it can be carried out in three hours rather than 5 days. However, the main limitation of COD is that it does not reveal whether the organic matter is biodegradable or non-biodegradable. Bio-refractory compounds need to be treated using physical and chemical methods rather than biological methods. The COD also includes inorganic oxidisable compounds.

The ratio of BOD/COD can be used to assess the biodegradability of organic compounds. For example, for municipal sewage the BOD/COD approximates to 0.68. Generally if the BOD/COD is greater than 0.4, then the substance is readily biodegradable.

Another common term used is 'total organic carbon' (TOC). The TOC is a non-specific test that measures the amount of carbon bound in organic material. All of the organics are oxidised to carbon dioxide and water.

Organics contribute to the effluent load of wastewater streams and is covered in Chapter 8. Organics also contribute to foaming in boilers and corrosion of steam piping. In cooling-water systems it serves as nutrients for micro-organisms.

2.3.7   Micro-organisms

Micro-organisms can be classified into two categories. First, those that pose a threat to human health (pathogens) and secondly, those which contribute to fouling and corrosion of equipment (but are non-pathogenic).

The most common human microbial pathogens found in water are enteric in origin. That means that enteric pathogens enter the environment in the faeces of infected hosts and therefore these are commonly found in sewage-contaminated water. Enteric pathogens can be viruses, bacteria, protozoa and Helminths. There are other pathogens that are not related to sewage contamination such as *Legionella* sp. *Legionella Pneumophila* bacteria found in cooling water systems is responsible for a type of pneumonia known as Legionnaires' disease.

2.3.7.1   Viruses

Viruses are the smallest of the pathogens found in water with diameters ranging from 20 nanometers (nm = 10^{-9} meters) to 85 nm [11]. In comparison a human red blood cell averages 7600 nm in diameter. The viruses associated with faeces are known as enteric viruses and are more than 100 entities. They can cause poliomyelitis, hepatitis and gastroenteritis to name a few diseases. The infectious dose can be as low as 10 viral particles or less. Many viruses are resistant to standard methods of disinfection. Viruses that infect bacterial cells are called bacteriophages. Bacteriophages are more abundant in wastewaters than viruses and for this reason the presence of bacteriophages are used as a proxy for virus concentrations with the results available in 5 days.

2.3.7.2   Bacteria

Bacteria are the most common and numerous of the microbial organisms in water. They can be pathogenic such as *Shigella dysenteriae*, *Campylobacter jejuni* and *Salmonella* or non-pathogenic such as *Psuedomonas*. Infectious dose can vary from as low as a few hundred cells for *Shigella dysenteriae*, *Campylobacter* to greater than a million in the case of other bacteria. *Salmonella* and *Campylobacter* bacteria are responsible for causing much of the food poisoning in the world. Figure 2.3 shows a scanning electron micrograph of *Salmonella* bacteria invading cultured human cells.

The presence of indicator organisms such as *E. coli* sp. or thermotolerant coliforms and/or total coliforms in drinking water indicates its faecal contamination. In cooling water systems, apart from the health risks caused by *Legionella*, bacteria such as *Psuedomonas* causes fouling and acts as biofilm

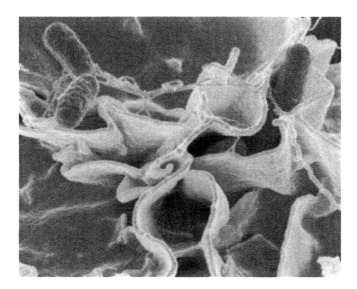

Figure 2.3   Scanning electron micrograph of Salmonella bacteria
Courtesy of Illinois department of Public Health.

for under deposit corrosion to occur. For this reason, total bacterial counts are closely monitored to minimise fouling. Bacteria that cause corrosion of metals are referred to as 'microbiologically influenced corrosion' (MIC). Both Sulphate Reducing Bacteria (SRB) such as *Desulfovibrio* sp. *and Clostridium* sp. belong to the MIC group. The SRB uses the conversion of iron-to-iron oxide to create energy. *Clostridium* sp. excretes hydrogen ions, which react with water to form strong organic acids. The common result is pits in the surface of the metal that are hidden under tubercles of iron oxide. Colonies can attack all types of ferrous metals including iron, mild steel, galvanised and stainless steel.

There are, however, beneficial uses of micro-organisms in sewage and industrial wastewater treatment plants. For example, bacteria can break down steel mill coke ovens gas contaminants such as cyanide, phenols and thiocyanate to undetectable levels.

Finally, floc-forming bacteria metabolise organic compounds in biological wastewater treatment plants. These aspects are discussed in Chapter 8.

Bacterial colonies are quantified as 'Colony Forming Units' (CFU/mL). It is a measure of the total viable plate count of the bacterial population.

2.3.7.3   Protozoa

These are unicellular organisms. Amoeba is the best example of this class of organisms. In water, wastewater systems organisms of interest are enteric pathogenic protozoa such as *Entamoeba histolytica*, *Giardia intestinalis* and *Cryptosporidium parvum*. All human protozoans are more infectious than the pathogenic bacteria. They are associated with faecal runoff. Filters with pore sizes smaller than 1 μm remove the oocysts.

2.3.7.4   Algae

These are both unicellular and multicellular organisms comprising of bacteria (blue green algae) and plants without roots, leaves or flowers. A characteristic of algae is their ability for photosynthesis. In cooling water systems they block nozzles.

2.3.7.5   Helminths

Helminths are parasitic worms and nematodes such as tapeworms. Other Helminth parasites of concern in reclaimed water are hook worm, round worm to name a few.

2.3.7.6   Fungi

Moulds are a class of fungi. Fungi produces spores. In cooling water systems mould causes wood rot.

2.3.8   Heavy Metals

These are commonly found in effluent streams of electroplating, textile dyeing, steel, petrochemical and other industrial plants. Whilst chromium is the most widely used heavy metal, cadmium, lead and mercury constitute the most toxic metals to humans. They have an attraction towards sulphur compounds found in enzymes. For this reason heavy metal concentrations are regulated by the EPA and water utilities.

2.3.8.1   Chromium

Chromium is a naturally occurring element found in rocks, animals, plants, soil and in volcanic dust and gases. Chromium is present in the environment in several different forms. The most common forms are as the element chromium(0), chromium(Cr^{3+}) and chromium(Cr^{6+}). No taste or odour is associated with chromium compounds. Chromium(Cr^{3+}) occurs naturally in the environment and is an essential nutrient. Chromium(Cr^{6+}) and chromium(0) are generally produced by industrial processes.

Chromium(Cr^{6+}) and chromium(Cr^{3+}) are used for chrome plating, dyes and pigments, leather tanning and wood preserving.

Whilst chromium(Cr^{3+}) is an essential micronutrient, contact with chromium(Cr^{6+}) can cause skin ulcers and ingesting large amounts of chromium(Cr^{6+}) can cause stomach upsets and ulcers, convulsions, kidney and liver damage, and even death.

Various metal finishing processes contribute chromium to the wastewater. Among these are chromium plating, chromating, bright dipping, chromic acid anodizing and chromium stripping.

The United States EPA has set a limit of $100\,\mu g$ for chromium (Cr^{3+}) and chromium (Cr^{6+})) per L of drinking water [12].

For methods of removal refer to Chapter 8.

2.3.8.2   Cadmium

Cadmium is a relatively rare element in the earth's crust and found in con-centrations of $0.2\,mg/kg$ [13]. It is usually found as a mineral combined with other elements such as oxygen (cadmium oxide), chlorine (cadmium chloride) or sulphur (cadmium sulphate, cadmium sulphide). The major use of cadmium is in the manufacture of nickel-cadmium batteries, electronic components, nuclear reactors, pigments, plastic stabilizers and in metal plating operations. Drinking water contains very low concentrations of cad-mium in the order of $0.1\,\mu g/L$. Polluted well waste may contain as much as $25\,\mu g/L$ [13]. Exposure to cadmium happens mostly in the workplace where cadmium products are made. The general population is exposed from breathing cigarette smoke or eating cadmium-contaminated foods.

Cadmium and its compounds are classified as carcinogens. They damage the lungs, can cause kidney disease and may irritate the digestive tract. Eating food or drinking water with very high levels severely irritates the stomach, leading to vomiting and diarrhea. Long-term exposure to lower levels of cadmium in air, food or water leads to a build-up of cadmium in the kidneys and possible kidney disease. Other long-term effects are lung damage and fragile bones.

The EPA has set a limit of five parts of cadmium per billion parts of drinking water (5 ppb). EPA does not allow cadmium in pesticides [12].

The United States Food and Drug Administration (FDA) limits the amount of cadmium in food colours to 15 parts per million (15 ppm).

2.3.8.3   Lead

Lead is a naturally occurring bluish-grey metal found in small amounts in the earth's crust. Lead can be found in all parts of the environment. Much of it comes from human activities including burning fossil fuels, mining and manufacturing.

Lead has many different uses. It is used in the production of batteries, ammunition, metal products (solder and pipes) and devices to shield X-rays. Because of health concerns, lead from gasoline, paints and ceramic products, caulking and pipe solder has been dramatically reduced in recent years.

Lead concentrations in drinking water range from $1\,\mu g/L$ to $60\,\mu g/L$ [14]. The effects of lead are the same whether it enters the body through breathing or swallowing. Lead can affect almost every organ and system in the body. The main target for lead toxicity is the nervous system. Exposure to high lead levels can severely damage the brain and kidneys in adults or children and ultimately cause death. Lead is not a carcinogen.

The EPA limits lead in drinking water to $15\,\mu g/L$.

2.3.8.4 Mercury

Mercury is a naturally occurring metal and exists in three oxidation states as the element, metallic mercury (+) and mercury (2+). The metallic mercury is a shiny, silver-white, odourless liquid. If heated, it is a colourless, odourless gas.

Mercury also combines with oxygen, sulphur chlorine to form inorganic salts and with carbon to make organic mercury compounds. The most common organic compound is methylmercury, is produced mainly by microscopic organisms in the water and soil. Metallic mercury is used to produce chlorine gas and caustic soda, and is also used in thermometers, dental fillings and batteries.

The nervous system is very sensitive to all forms of mercury. Methylmercury and metallic mercury vapours are more harmful than other forms, because more mercury in these forms reaches the brain. Exposure to high levels of metallic, inorganic or organic mercury can permanently damage the brain, kidneys and developing foetus. Effects on brain functioning may result in irritability, shyness, tremors, changes in vision or hearing and memory problems.

Short-term exposure to high levels of metallic mercury vapours may cause effects including lung damage, nausea, vomiting, diarrhea, increases in blood pressure or heart rate, skin rashes and eye irritation.

The EPA has set a limit of 2 µg/L in drinking water (2 ppb).

2.3.9 Radionuclides

Drinking water contains naturally occurring radionuclides such as radium-226 and uranium-238. Wastewater streams discharged from hospital laboratories and nuclear power and research facilities contain radionuclides such as uranium and radium-226. The naturally occurring form of Uranium is as the Uranyl ion UO_2^{2+}. Uranium, while it may be radioactive, is actually more serious as a toxin to the kidney. At high enough levels, it can cause permanent kidney damage. The current US EPA standard for Uranium in drinking water is 0.1 mg/L [15]. Similarly other national drinking water standards provide specific limits to be observed.

References

[1] Department of Water Affairs & Forestry. *South African Water Quality Guidelines*. 2nd edition, Vol. 1. *Domestic Water Use*. 1996.

[2] Illawarra Public Health Unit. *What is Hydrogen Sulphide*. www.shoalhaven.nsw.gov.au/Environment/HydrogenSulphide.pdf.

[3] National Health and Medical Research Council, Agricultural Resource Management Council of Australia and New Zealand. *Australian Drinking Water Quality Guidelines – Summary*. 1996.

[4] National Health and Medical Research Council, Agricultural Resource Management Council of Australia and New Zealand. *Guidelines for Sewerage Systems – Reclaimed Water*. February 2000.

[5] Judd S. and Jefferson B., 1st edition, *Membranes for Industrial Recovery and Reuse*. Elsevier Ltd, UK. 2003.

[6] Arthur H.T. *The Right Metal For Heat Exchanger Tubes*, Chemical Engineering, Mc GrawHill Inc. New York. 1990.

[7] Bennett P.B., *Fundamentals of Cooling Water Treatment*. Booklet 11–431. Calgon Corporation, Pittsburg.

[8] Bauder T.A., Waskom R.M. and Davis J.G. *Irrigation Water Quality Criteria*. Colorado State University Cooperative Extension Service. December 2004.

[9] Department of Water Affairs & Forestry. *South African Water Quality Guidelines* 2nd edition, Vol. 3. *Industrial Water Use*. 1996.

[10] Environment Protection Division. British Columbia. *Ambient Water Quality Criteria for Colour in British Columbia*. www.env.gov.bc.ca/wat/wq/BCguidelines/colour/colour-08.htm.

[11] BioVir Laboratories Inc. Enteric Virus. Benicia, CA. www.biovir.com.

[12] Agency for Toxic Substances and Disease Registry. ToxFAQs for Chromium. http://www.atsdr.cdc.gov/tfacts7.html.

[13] World Health Organisation. *6_3 Cadmium*. www.euro.who.int/document/aiq.

[14] World Health Organisation. *6_7 Lead*. www.euro.who.int/document/aiq.

[15] Johnson G., Stowell L. and Monroe M. *VSEP Treatment of RO Reject from Brackish Well Water*. New Logic Research Inc. Proceedings of the 2006 El Paso Desalination Conference. El Paso, Texas. 15–17 March 2006.

Chapter 3

Saving Water: Step by Step

3.1 Developing a Sustainable Water Management Plan

To save water an organisation requires a plan. A successful water management plan (WMP) needs to incorporate *technical approaches and a systems approach*. This is the key to achieving **sustainable** water conservation within a facility.

A plan works best when it places *sustainable water management* in the context of an organisation's overall approach to social and environmental responsibility. Without a systems approach, gains achieved through technical improvements are lost when management finds other priorities, or when a champion is promoted or leaves the organisation. It is only when costs increase again that another campaign is launched and old ground is revisited as illustrated in Figure 3.1. To avoid this 'roller coaster' cycle, management must ensure that there is a *continuous improvement strategy*.

A successful WMP

- is part of a well-formulated water policy that aligns with the goals of the corporate plan and regulatory requirements (**leadership**)
- identifies and assigns responsibility to a person (**accountability**)
- sets targets for water use efficiency and minimises pollution to reduce the overall environmental footprint (**plans**)
- integrates water efficiency into the operations of the organisation
- addresses behavioural changes among employees
- drives the development of water usage reporting systems and processes for collection of data
- specifies the costs and benefits to the company of sustainable water management and
- drives changes in how the organisation interacts with contractors and suppliers.

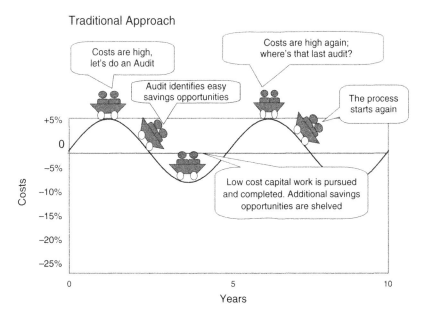

Figure 3.1   Traditional approach to water management

Courtesy of Energetics Pty Ltd.

The benefits of having a well-formulated WMP are as follows:

- Alignment with corporate strategy

  Water conservation can be brought into alignment with corporate strategy by showing the linkage between water conservation, environmental performance and business improvement. Once this linkage is clearly shown it will provide a platform for the organisation to allocate sufficient resources to carry out the WMP. The alignment with corporate strategy will then lead to systems being embedded to realise long-term sustainable savings in water and other associated resource input costs while improving environmental performance of the organisation. Figure 3.2 shows this relationship.

  For example, the steel industry is a highly competitive industry. To be competitive in the global market a steel manufacturer needs to operate in the lowest cost quartile. When this is the corporate strategy, then reducing all input costs including water needs to be a high priority for the company.

- Cost savings

  A reduction in water consumption will reduce both water usage costs and discharge costs as well as reduce other associated costs such as natural gas costs, electricity and chemical treatment costs. Improved

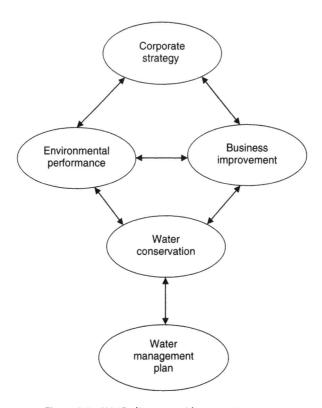

Figure 3.2   WMP alignment with corporate strategy

water efficiency in a manufacturing plant can also provide productivity and product quality improvements.

- Reduced organisational water footprint
  A water management plan will result in making significant reductions in water usage if an organisational perspective is maintained. All inputs and outputs need to be analysed.

Case Study: BlueScope Steel

BlueScope Steel Port Kembla Steel Works is Australia's largest steel manufacturing facility. It consumes $35\,000\,m^3$/day (9.25 US Mgal/day) of freshwater and $850\,000\,m^3$/day (224.5 US Mgal/day) of saltwater. Over a ten year period BlueScope Steel has improved water use efficiency per ton of steel from $5.5\,m^3$/T to $2.69\,m^3$/T making it one of the world's most water efficient steel makers [1].

- Good public relations

  A successful water conservation programme can be promoted internally to the organisation. This will increase uptake by senior management to allocate funds for capital- intensive water conservation initiatives. By publicising the achievements externally, the organisation will come to be regarded as an environmentally committed good corporate citizen.
- Brings about behavioural changes through involvement of employees, contractors and suppliers

  A successful WMP involves all employees, relevant contractors and suppliers. Therefore a WMP should contain sufficient initiatives such as workshops and seminars to increase employee awareness as well as to get their participation in identifying the improvement opportunities. This ultimately leads to positive cultural and behavioural change within the organisation. Employees can be motivated through financial and non-financial incentives. Contractors may be evaluated using additional performance criteria, which will align their activities with the organisation's water conservation efforts.

  For example, cooling tower water treatment contracts are typically awarded based on lowest cost, technical competence and on-site availability. Minimising water usage is rarely a requirement. A new contract could explicitly state the need to ensure that cooling tower water usage is minimised whilst maintaining the other criteria. This in turn reduces the chemicals required to treat the water and results in a cost saving.
- Reassessment of redundant process steps

  Often these initiatives lead to reassessment and review of processes. Often processes are carried out through force of habit and rarely questioned whether they are valid or not. A WMP will lead to reassessment of the process steps and help to identify redundant steps.
- A reference document for obtaining subsidies

  Many governments and municipal authorities provide funding to implement water conservation within the business community. A well-developed WMP can be an excellent document to cite when applying for funds from these agencies as it shows the organisation's commitment and structured approach to water conservation.
- Linked to senior management performance

  The WMP is a transparent document to allocate responsibilities and water usage reduction targets to departments and senior managers. The achievement of WMP targets can be linked to senior management remuneration packages. Often whilst the corporate strategy may have water conservation as a key driver for the organisation, middle managers may not be fully committed to achieving those drivers. By linking senior management remuneration to achieving those targets, the organisation is ensuring that the goals will be achieved.

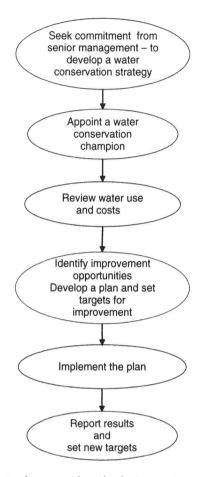

Figure 3.3   A step-by-step guide to developing a water management plan

Figure 3.3 gives a step-by-step guide to establishing a successful Water Management Plan.

3.2   Step 1: Seek Senior Management Commitment

Senior management commitment and leadership is fundamental to the success of a sustainable water conservation programme. Therefore, management needs to do the following:

- Develop a water conservation policy that includes clear water reduction targets, not just cost savings. The policy can be developed in terms of reducing overall consumption compared to a base year or by reducing it to below an industry benchmark. If there is lack of knowledge on how to set clear targets, then this could be done after the saving potential has been identified.

An example of a water conservation policy

This organisation is committed to reducing its water consumption. By 2010 we will reduce our potable water consumption from $4\,m^3$/ton to $2.0\,m^3$/ton. This will increase efficiency, cut costs and enable us to be globally competitive whilst making a positive contribution towards the environment.

- Appoint a company-wide water conservation manager.
- Allocate sufficient resources.
- Bring about cultural change within the organisation by communicating the policy internally to all employees, contractors and external stakeholders.

Begin with a simple positive message to all employees about the importance the business is placing on the programme, why the organisation is doing it and explain their role. The message needs to emphasise the need for employee involvement and support. By doing so, the water conservation programme will become sustainable, as attitudes to water change from a *consumable* to a *business resource*.

3.3   Step 2: Appoint A Water Conservation Manager

The secret to a successful programme is having a champion within the organisation with responsibility and authority to transform the policy into a workable plan.

The manager needs to be given an achievable target, adequate funding and support. The target to be linked to the manager's remuneration package. This way the manager's personal goals are in alignment with organisational requirements.

3.3.1   Responsibilities of the Water Conservation Manager

- Form a water conservation working group drawn from different divisions and units.
- Carry out a management review of business practices – otherwise known as a management diagnostic. The objective is to identify the

system barriers or lack of systems that are inherent within an organisation, which prevents the organisation from developing a successful sustainable water management programme.

- Review and evaluate the organisation's existing (or previous) water conservation programmes. Note areas that were successful and areas that were not effective.
- Establish a budget and secure the necessary funding.
- If site-wide audits are not being done, initiate water audits and provide assistance to auditors. Investigate water reuse opportunities.
- Create the water conservation action plan. The plan should include the goals of the programme as well as the details for implementing specific water conservation measures (both technical and management improvements).
- Establish the process by which the water conservation plan will be documented and evaluated.
- Implement the water conservation programme. Begin with the lowest cost conservation measures that have quick paybacks. This will give the programme a boost through quick wins and motivate others.
- Recognise and reward employees whose suggestions and actions achieve water reductions and publicise those achievements both internally and externally; for example through newsletters.
- Continually monitor water use through water meters (Worksheet 1).
- Regularly report water conservation progress against targets to senior management. Each initiative needs to have a goal, when it would be implemented and what was achieved and issues encountered. Review the plan annually, analyse the gap against goals and commitments and make changes to create additional water reductions.

3.4   Step 3: Gather Baseline Data and Review Usage

The first task for the water management champion is to collate information pertaining to water usage and costs to develop a water balance for the site.

Lack of usage data creates the myth that water is a fixed cost. This is not the case. Water is a variable cost and it empowers the user to reduce its consumption.

Table 3.1 gives the type of data that should be collected to gain an understanding of a water balance.

Actions

1. From the water and sewer bills for the last 2 years develop a spreadsheet of water consumption, charges for water, wastewater and trade

Table 3.1 Gathering existing data

Type of data	Description
Water charges (fixed plus usage)	Collect water bills for the previous 2 years. If not available contact your local water authority.
On-site treatment costs (fixed plus variable)	Chemical treatment (or other costs) to purify the water further, such as using reverse osmosis, ion exchange.
Sewer usage charges and trade waste charges	Wastewater bills for the previous 2 years. If not available contact your local water authority.
Effluent pre-treatment costs and quality	Chemical treatment costs for dissolved air flotation and other pre-treatment systems and discharge wastewater quality.
Sewer usage discharge factor	Per cent of effluent discharged to the sewer as per cent of inlet water.
Effluent- and sludge-removed off-site	Waste disposal costs and frequency of sludge removal.
Site plans	Location of meters, hydraulic plans, distribution system.
Number of main and sub-meters	List the number of main meters and sub-meters, type and size and if they are capable of being remotely logged.
Unit operations description	Process flow and pipe/process technical drawings.
Production data or metrics used by the organisation	In retail outlets use number of transactions. For service establishments like hotels; laundries; hospitals; canteens; military establishments; school records of meals served; rooms occupied; patients or kilogram of dry linen.
Number of employees and contractors	Useful to generate benchmark information.
Labour	Labour costs to operate wastewater treatment plant.
Energy costs	Gas and electricity costs.
Major water using equipment	A detailed inventory of major water using equipment including their water usage rates etc.

waste. Use Worksheets 2 and 3 in the Appendix or request it from the water utility. Sydney Water' Every Drop Counts Business Program provides this service as shown in Figure 3.4.

2. Chart water consumption per month or billing period and insert production data. Generate raw benchmarks as shown in Figure 3.5.
3. Link water usage to other costs such as chemicals to find true cost of water for your business.
4. Begin monitoring water usage through main meters and sub-meters to gather baseline data.

Figure 3.5 shows a water consumption profile in an office building with target benchmarks.

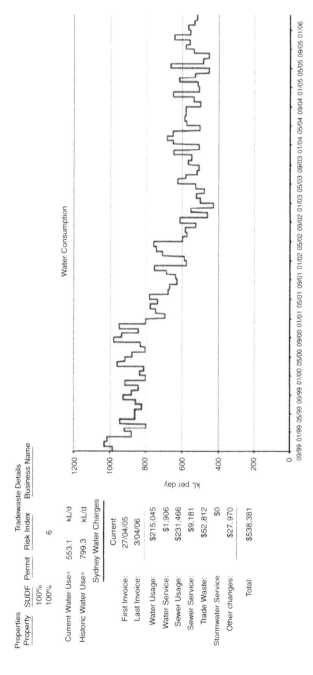

Figure 3.4 Sydney Water customer usage profile

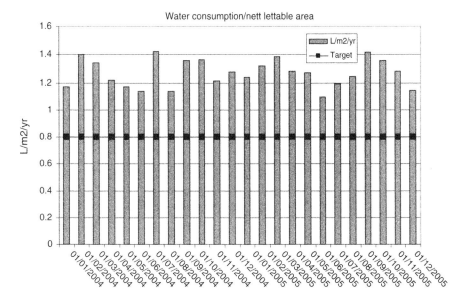

Figure 3.5   Water consumption profile in an office building per square metre

3.5   Step 4: Identify Improvement Opportunities

To identify improvement opportunities it is necessary to carryout an assessment of management systems and physical systems. It will become apparent that the two need to be addressed simultaneously for long-term gains.

3.5.1   Carry Out an Assessment of Management Systems

A self-assessment of management systems will identify system barriers to water conservation as well as help to ensure that systems and processes are developed to maintain the gains over the long term. A structured systems approach eliminates the reliance on the motivation of one single person.

The objectives of a management diagnostic are to identify the following:

- links water with the organisation's value drivers such as raw material inputs, production processes, distribution, product use and discharge
- identifies the key system barriers, opportunities and risks associated with developing a sustainable water management plan
- allocates responsibilities that go beyond the immediate project team and
- ensures that water-related actions, opportunities and risks are tracked and managed effectively using a continual improvement process.

3.5.1.1   One-2-Five Water® – Management Diagnostic System
There are many different types of management diagnostic systems such as the management diagnostic developed by the Global Environmental

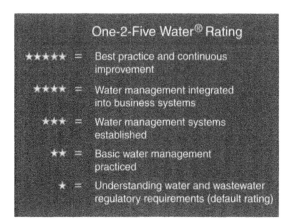

Figure 3.6   One-2-Five Water® Rating

Courtesy of Energetics Pty Ltd.

Management Initiative and known as *Connecting Drops Toward Creative Water Strategies* [2] that can be accessed from their website: www.gemi.org. We will focus on One-2-Five Water® [3].

One-2-Five Water® was developed by the environmental consultancy Energetics with the assistance of Sydney Water. It is based on a five-star rating system as shown in Figure 3.6.

The One-2-Five Water® diagnostic tool allows a team of management and operational staff – in an hour or so – to conduct a self-diagnostic. At the end of the session the tool provides

- a star rating
- a percentage development required to reach the next level
- five critical actions for the organisation to carry out to reach the next level of development.
- allocates responsibilities and completion dates.

Sydney Water through its Every Drop Counts Business Program has facilitated over 230 such management diagnostics. These are shown in Figure 3.7.

The rating analysis looks at the ten building blocks necessary for water management. These are shown in Figure 3.8.

A One-2-Five Water® diagnostic results are shown in Figure 3.9. For each critical action it shows the action required.

3.5.2   Technical Assessment

The management diagnostic will indicate if the site requires a water audit if one has not been done recently. The water audit will identify where the water is used as well as show the amount of water used by critical equipment,

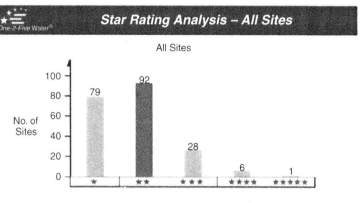

Figure 3.7   Distribution of One-2-Five Water® Star Ratings
Courtesy of Energetics Pty Ltd.

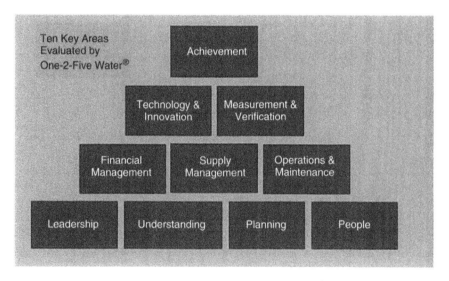

Figure 3.8   Elements of One-2-Five Water®
Courtesy of Energetics Pty Ltd.

identify leakage, help to develop a water balance and provide a benchmark for the individual site.

3.5.2.1   How Detailed Should the Water Audit Be?
This decision should be made upon the following considerations:

- Average daily water consumption on site. A walk through audit is adequate if the water usage is less than $50 \, m^3$/day (<13 000 US gal./day).

Diagnostic Results

Element	1 Star	2 Star	3 Star	4 Star	5 Star	Critical Action Items
1.1 Demonstrated corporate commitment	–	●	–	–	–	Critical
2.1 Understanding of performance and opportunities	–	✎	✎	–	–	–
3.1 Targets, performance indicators (KPI) and motivation	–	✎	–	–	–	–
3.2 Plans	–	✎	–	–	✎	–
4.1 Accountabilities	●	–	–	–	–	–
4.2 Awareness and training	–	–	✎	–	–	Critical
4.3 Resourcing	–	–	✎	–	–	–
5.1 Criteria/Budgets for capital expenditure (CAPEX)	–	–	✎	–	–	–
5.2 Operating budgets	–	✎	–	–	–	–
6.1 Water supply, Quality & Reliability	–	–	–	–	–	–
6.2 Compliance with legal and other requirements	–	✎	–	✎	–	–
7.1 Operating procedures	–	–	–	–	–	–
7.2 Maintenance procedures	–	✎	✎	–	–	–
8.1 Efficiency of existing plant design	–	✎	✎	–	–	–
8.2 Innovation and new technology	–	–	–	–	–	–
9.1 Metering and monitoring	–	●	–	–	–	–
9.2 Reporting, feedback and control systems	–	●	–	–	–	Critical
9.3 Documentation and records	●	–	–	–	–	Critical
10.1 Water cost performance in the past 12 months	–	✎	–	–	–	Critical

Overall Ranking: 2 Star   % Achievement: 38%   % Achievement to reach next level: +8%

Figure 3.9   One-2-Five Water® Results

Courtesy of Energetics Pty Ltd.

A detailed audit if the water usage is greater than 50 kL/day (>13 000 US gal./day).

- Is it a regulatory requirement? In Sydney, the NSW Government has mandated that businesses using over 50 ML/yr (over 200 businesses) and local government councils develop water-saving action plans. This requires that water audits to be done in these premises.
- Is the majority of the water used in one equipment? – such as in commercial laundries where almost 80% of the water is used in the tunnel washers. Then the focus needs to be on water reuse from the tunnel washer or replacing the washer if it makes financial sense.
- Likelihood of identifying cost-effective opportunities to save water based on a comparison with industry benchmarks or industry best practice if readily available. See Section 3.5.2.2.
- Multiple similar sites (a single detailed audit can be used to benchmark other sites).
- Other potential savings in energy or materials.
- Availability of budget for funding water audits or other investigations.
- Cost of water audits and other investigations.
- Age of the infrastructure – in many instances, leakage from underground pipes, urinals and taps can account for 20–30% of the usage in facilities such as in hospitals, prisons, hotels and so on. Avoiding water wastage should be the first priority.

3.5.2.2   Estimating Water-Saving Potential

Various state government agencies have over the years carried out studies to estimate the water-saving potential from the business sector. The Metropolitan Water District of Southern California found that for the commercial sector the average potential water savings per year is about 20% of average consumption and the institutional sector is about 19% [4].

Another study conducted by the US Environmental Protection Agency (EPA, 1997) found that commercial water-use volume might be cost effectively reduced by approximately 23% [5].

Two US studies conducted by the Pacific Institute and the New Mexico Drought Task Force puts this figure at 40% [6, 7]. In some sectors this can reach as high as 75% through water reuse and recycling – such as in oil refining.

According to the UK Environment Agency, water reduction of 40% in commercial buildings and up to 90% in industrial sites can be achieved [8].

Sydney Water's Every Drop Counts Business Program has found that the average water savings based on water audits in commercial and institutional sectors range from 20 to 40%. Similar to the findings of the Pacific Institute study, industrial facilities can save from 20 to 80% through water efficiency, water reuse and recycling.

A general rule of thumb

- If no water-saving measures have so far been implemented, savings could be 20–50% or more of water-related costs.

- If some water-saving projects have been implemented but not applied using a systematic approach, the potential savings could be at least 20% of water-related costs.
- Water-saving potential needs to be considered using the resource minimisation hierarchy of **Avoidance, Reduce, Reuse and Recycle**. In some cases, through avoidance 100% of water savings can be achieved.

Table 3.2 shows the potential water savings in the commercial and institutional sectors from a large number of audits carried out in the commercial, hospitality, government and institutional sectors in the United States.

Table 3.3 shows potential water-saving projects in industrial plants.

Another way of estimating water-saving potential is to compare against industry benchmarks. For instance in the United Kingdom, public buildings are required to reach a best practice water usage level of $6.4\,m^3$/person/yr [10]. Other examples of benchmarks are given in Table 3.4 and in Chapters 10–15.

The percentage reduction values and industry benchmarks whilst useful as a guide are only to be used for comparative purposes. In many instances the age of the equipment, layout, production method and other local factors may have a bearing on the target. For instance, the UK brewing industry produces more draught beer which consumes less water and energy than canned beer.

3.5.2.3 Complying with Regulatory Standards

Increasingly regulatory authorities are mandating that water-efficient fixtures be offered for sale and installed in buildings. For instance, from July 2006

Table 3.2 Potential water savings from on-site water audits [9]

Type of Business	Number of Audits	Average (%)
Car wash	12	27
Church non-profit	19	31
Communications and research	10	18
Corrections	2	14
Eating & drinking	102	27
Education	168	20
Healthcare	90	25
Hospitality	222	22
Hotels & accommodations	120	17
Landscape irrigation	6	26
Laundries	22	15
Meeting and recreation	20	27
Military	1	9
Offices	19	28
Sales	56	27
Services	58	30
Transportation and fuels	24	31
Vehicle dealers and services	12	17
Total sites	963	

Table 3.3 Water-Saving Potential in Industrial Plants

Area	Project	Potential saving
All	Minimise leaks	10–30%
Amenities	Water-efficient fixtures	3–10%
Utilities	Conversion from once-through cooling to closed-loop recycle	90%
Utilities	Conversion from once-through cooling to open evaporative cooling	Upto 60%
Utilities	Recovery of steam condensate	Upto 75–80% (depends on sector)
Utilities /Process	Reuse of effluent in cooling water systems and in process	Upto 75% (depends on water quality)
Process	Clean in Place – Automation, Optimisation	60% [8]
Process	Air rinsing of containers	100%
Process	Air thawing of meat	100%
Process	Counter-current rinsing	30–40%
Process	Spray jet upgrades	20% [8]

Table 3.4 Water usage/metrics

Category	Australian, UK and European metrics	US metrics
Oil refining total water usage/unit	0.1–4.5 [11] m³/ton of crude	65–90 gal./barrel of crude
Office buildings	30–135 [12] L/employee/day 0.8 m³/m²/yr (Sydney) 9.3 m³ per person per year (UK)	8–20 [13] gal./employee/day
Hospitals	1.17–1.66 [14] m³/m²/yr	130–250 [13] gal./bed/day
Brewing	2.5–7 L/L of beer	2.5–7 gal./gal. of beer
Prisons	92.4–115.3 [14] m³/prisoner/yr	80–150 [13] gal./prisoner/day

in Australia, only water-efficient fixtures can be sold under the Australian Water Efficient and Labelling and Standards (WELS) scheme. This means it will be mandatory for showers; clothes washing machines; dishwashers; toilet equipment; urinal equipment; and tap equipment intended for use over kitchen sinks, bathroom basins, laundry tubs or ablution troughs to carry a WELS Water Rating label when they are offered for sale (*AS/NZS6400:2005 Water-efficient products – Rating and labelling* (Schedule 1 to the *Water Efficiency Labelling and Standards Determination 2005*)).

Based on a six-star scheme, the more stars the more water efficient a fixture is. Figure 3.10 shows a three-star water rating label for showers.

In the United States, the Energy policy Act of 1992 requires that Federal agencies not later than 1 January 2005 implement in Federal buildings all

Figure 3.10   WELS Rating for showers

Courtesy of Department of Environment and Heritage, Australia.

energy and water conservation measures with a payback period of less than 10 years.

In the United Kingdom, under the Water Smart project, all government office buildings are required to monitor water usage, report it on the World Wide Web and make progress towards the target. For more details, refer to Chapter 11.

3.5.2.4   Carrying out a Water Audit

Establish base flow.   Base flow is the water that is consumed by a site during non- working hours usually shown by a constant consumption during these hours. Base flow occurs due to water-using equipment left on after hours such as cooling towers operating continuously, due to faulty maintenance or due to pipe leaks. The 2-year water consumption profile does not indicate if there are leakages occurring during non-working hours. To establish whether there are any leakages on site, connect a data logger to the main meter or to sub meters.

Figure 3.11 shows the value of real-time data logging to detect a 3 L/s base flow ($259\,m^3$/day).

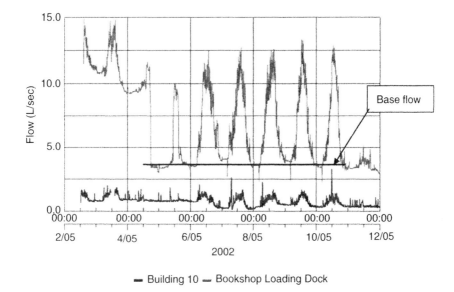

Figure 3.11 Real-time data logging to identify base flow

Table 3.5 Water losses and its cost

Base flow L/s	Base flow/day		Example	Cost* per annum
	m³/d	Thousand US gal./d		
0.5	43	11.4	25 mm (1 in.) Hose	A$37,668
1	86	23	Vacuum pump once-through usage	A$75,336
2	173	46	Underground leak to stormwater	A$151,548
4	345	91	Tank and cooling water overflow from cooling tower	A$302,220
4.6	400	106	Tank overflow	A$350,400

* Cost are based on water usage $1.20/m³ and wastewater discharge $1.20/m³.

Base flow can consist of tank and cooling tower overflows, hoses not shut or equipment using once-through water. Examples on how base flow can account for significant water wastage is shown in Table 3.5.

The case study below shows how a hospital was able to reduce their water costs by A$200 000 by merely monitoring the main meter and taking simple corrective action to eliminate base flow.

If dataloggers are not available (or if the meter cannot give a 4–20 mAmp signal), then either change the meter or take manual readings after working hours and the following day before work begins. For more details, refer to Chapter 4.

Case Study: Prince of Wales Hospital, Sydney, NSW, Australia

The Prince of Wales Hospital, through data logging of the main meter, identified a 4-litre-per-second leak amounting to 345 m³/day. This had gone unnoticed (since it only leaked at night) and was only detected due to the data logging of the main meter. Cost saving – A$230 000 per annum.

Sydney Water. *Leaks Waste More than Water. The Conserver*, Issue 9, August 2005.

The location of the meter can be further investigated through progressive shutdowns.

Verify flows. Fixtures such as taps, showerheads and toilets can waste a lot of water or may be in excess for the purpose. Chapter 10 discusses methods to reduce water wastage in amenities.

Figure 3.12 shows annual water losses from taps.

- From the hydraulic plan identify the equipment that uses more than 15% of the total water consumption and install sub-meters to these areas. This will help in developing the site water balance. Either read the meters manually or better install dataloggers. These sub-meters can be either permanent or temporary. If sub-meters are difficult to install then measure flows using an ultrasonic flowmeter. Refer to Chapter 4 for more information on meters, dataloggers and telemetry.

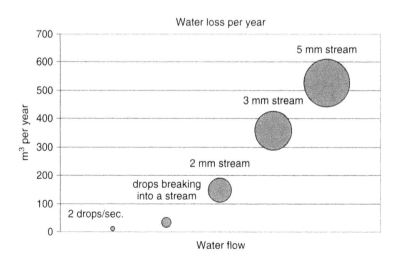

Figure 3.12 Water losses from taps – what it means

- Use manufacturer's data and compare the equipment's water use with the manufacturer's rated flow amounts. Some equipment may be using more water than the specified flow rates. If there is a significant difference, consider having a specialist review equipment operation and make adjustments to reduce water consumption.
- Estimate water use from knowledge of the process.
- Identify the quality of water as it travels through a unit or facility. Discharges from one area could be used as the supply water for a second use. Typical areas of interest are temperature, chemical constituents (including pH), total dissolved solids and/or conductivity.
- Observe visual leakage and note for immediate action. On a regular basis, thoroughly check the following areas:

  - restrooms, and shower facilities
  - kitchens, dishwashing facilities and food preparation areas
  - wash-down areas and janitor closets
  - water fountains
  - water lines and water-delivery devices
  - process plumbing, including tank overflow valves and
  - landscape irrigation systems.

3.5.2.5 Develop a Water Balance

A water balance needs to capture the main water-using areas. It should account for at least plus or minus 80% of the input flows.

The decision on how detailed the water balance should be would need to consider the following:

- likelihood of identifying cost-effective opportunities to save water
- cost
- calculation of the additional cost savings.

The higher the amount of water used, the greater the potential to identify additional water savings.

From the flow monitoring data we can assign flow rates to the major water using facilities. Figure 3.13 shows the water balance of a hotel. Any discrepancies are due to hidden losses, incorrect meter readings, incorrect assumptions or faulty meters.

The collected information can also be expressed as a pie chart. A pie chart as shown in Figure 3.14 is an effective visual means of showing the main water-using areas and can be used as a management reporting tool.

3.5.2.6 Identifying Other Opportunities to Reduce Water Use

Use the water balance to further identify opportunities to reduce water usage. Involve the operators of the equipment to generate other water-saving ideas.

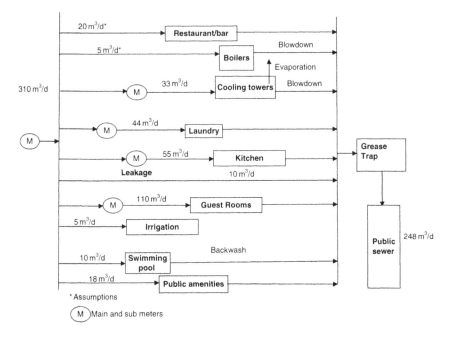

Figure 3.13　A Water Balance of a 300-Room Hotel

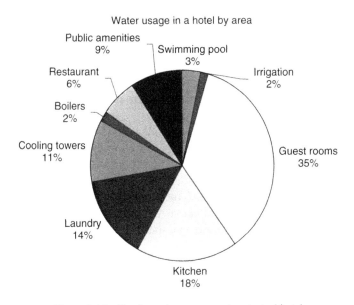

Figure 3.14　Pie chart of water usage in a typical hotel

In generating the options list use a checklist that includes the following:

1. Can the maintenance activities be improved? For an example, steam traps require a regular preventative maintenance programme. Other areas are missing jets in dishwashers, leaking pipes, flange joints, overflowing of cooling towers due to faulty float valves and so on.
2. Can water pressure be reduced?
3. Improve housekeeping. Use brooms to reduce hosing of kitchen and plant floors without compromising operational guidelines.
4. Is it possible to reduce the water usage in the process/activity? – For example, liquid ring vacuum pumps waste water if the sealing water is once through. A dry running pump eliminates the need for water. Or recycling the water reduces the water usage significantly.
5. Will a lower-quality water be sufficient? – For example, as cooling tower make up.
6. Can the water currently sent to the sewer be recovered? – As low grade water or after treatment to be used in non-potable applications.
7. Compare with industry benchmarks to ascertain water usage per activity if this data is available.
8. Do the hoses have automatic trigger guns?
9. What behavioural practices require changing? Changes to flushing of toilets when cleaning guest rooms. Retraining of kitchen staff to load the dishwasher to the maximum.

3.6 Step 5: Preparing the Plan Prioritising the Opportunities

From the gathered data calculate the potential water savings and associated cost savings – including heat recovered. Determine costs to implement water-saving actions. Where applicable obtain quotes from contractors and suppliers, including time required for implementation. If not already known, find out the typical capital expenditure hurdle rates used by your business and calculate return on capital. Or for simplicity use the payback method.

The recommendations can be prioritised as shown in Table 3.6.

The water-saving plans need to be specific, quantifiable and achievable. The goals need to include the following:

*   a volumetric reduction target as m^3/yr, or a percentage of water saved and/or a benchmark figure as m^3/unit of output (L water/L of milk, L/person/yr or L/area/yr)
*   a time frame for achievement and
*   an area (of the facility or unit) where the water savings will be realised.

Table 3.6 Tabulation of recommended actions

Priority	Technical risk
Requires immediate action (arrest leakage) Short payback	None. Maintenance activity.
Cost effective and practical. Generally accepted as payback within 2–3 years	Low technical risk. Does not require further investigations
Potentially viable. Payback above 3 years but less than 5 years.	Medium technical risk. Require further investigations.
Not cost effective. Payback above 5 years	Medium technical risk.
High technical risk and unknown payback.	No proven case studies. Require extensive investigations.
Measures not having a quantifiable payback (e.g.: installing a centralised water monitoring system, increasing employee awareness)	None.

The benchmark figure is preferred because it is a true measure of water efficiency. It allows for increasing water usage if production capacity increases by measuring water usage per unit of output. These goals can be revised with time.

Worked example

A 300-room hotel in Sydney uses $310\,m^3$/day of water. From the Best Practice Guidelines for Hotels (refer to Chapter 10) it has been estimated that a 300-room hotel with a cooling tower and laundry require only $150\,m^3$/day. The hotels current consumption is $310\,m^3$/day as per the water audit and water balance. Therefore there is the potential to save $160\,m^3$/day.

The opportunities are shown in Table 3.7 based on the water-use inventory (Worksheet 3).

Implement the prioritised measures and measure the reduction in water usage. Set a new benchmark. In the example the target benchmark is $150\,m^3$/day.

3.7 Step 6: Report the Results

By communicating water conservation achievements and new targets and challenges to senior management, staff, tenants and guests, contractors and suppliers and regulatory authorities there will be greater support for the programme from all making it easier justify projects and new ideas will be generated and gain recognition for the programme.

Table 3.7 Prioritisation of water saving measures

Water-saving measure	Number Of fixtures	Water Savings m³/day	Cost savings /yr A$	Cost to implement A$	Simple payback	Can this be done immediately?	Time required to make changes	Technical risk	Order of Priority for implementation
Fix leaks	5	20	17,500	$3000	2 months	Yes	5 days	None Maintenance activity	1
Guest rooms – retrofit showers with 9 L/min showers in guest rooms instead of 15 L/min	300	32.4	18,921	$36,000	1.9 years	Yes	14–30 days	None Proven technology	1
Retrofit taps with water-efficient aerators 6 L/min instead of 12 L/min	400	36	23,652	8,000	4	Yes	30 days	None	
Cooling tower float valve replacement	1	10	8,760	200	0.3 month	Yes	14 days	None Maintenance activity	1
Kitchen – replace pre-rinse spray valves	3	11.6	13,033	1,200	1.0 month	Yes	1 day	None	1
Kitchen – replace wok stoves with waterless woks	2	10	8,760	10,000	1.1 year	Yes	7 days	Proven product	1

(Continued)

Table 3.7 (*Continued*)

Water-saving measure	Number Of fixtures	Water Savings m³/day	Cost savings /yr A$	Cost to implement A$	Simple payback	Can this be done immediately?	Time required to make changes	Technical risk	Order of Priority for implementation
Reuse of laundry effluent in laundry	1	30	26,280	$65,000	2.5 years	No	Shutdown	Medium risk Further investigations required	2
Guest rooms – replace single-flush toilets with dual-flush toilets	300	11	3,000	$120,000	40 years	No	12 months	None	3 Not cost effective
Employee training programme		TBD	TBD	3,000		No	3 months	None	2
Employee training programme		TBD	TBD	3,000		No	3 months	None	2
Install dataloggers to sub-meters	5	TBD	TBD	6,000		Yes	1 month	None	2
Total		161	119,906	252,400					

1 = First priority, 2 = Medium priority, 3 = Not a priority
Discount Rate   10%
Net Present Value $ 227,536
Internal Rate of Return 70%
(Please refer to Chapter 9 for an explanation on Net present value and IRR)

This can be done by

i)   A newsletter
ii)   Media release
iii)   Bulletin boards with graphs showing the reduction in water usage
iv)   Hotel room cards
v)   Pay-check inserts
vi)   Staff meetings
vii) Employee recognition and incentive programmes.

3.8   Conclusion

A holistic organisational approach is required to achieve sustainable reduction in water usage. It starts with the organisational commitment; an appointment of a competent person as a water manager; and water management needs to be done in a systematic manner. A systems approach ensures continuous improvement, eliminates ad hoc approaches and keeps the whole organisation focused on achieving the policy objectives and informed of the progress towards targets.

References

[1]   Sydney Water. Every Drop Counts at *BlueScope Steel*. The Conserver Issue 10. May 2006.
[2]   Global Management Initiative. *Connecting Drops Toward Creative Water Strategies A Water Sustainability Tool*. www.gemi.org.
[3]   www.energetics.com.au.
[4]   ERI Services. *Metropolitan Water District of Southern California: Commercial, Industrial, and Institutional Water Conservation Program, 1991–1996*. 1997.
[5]   U.S. Environmental Protection Agency. *Study of Potential Water Efficiency Improvements in Commercial Businesses*. 1997.
[6]   Pacific Institute. *Waste Not Want Not: The Potential for Urban Water Conservation in California*. 2000.
[7]   New Mexico Drought Task Force. *Industrial Commercial and Institutional Water Conservation*. June 2004.
[8]   Environment Agency UK. *Water Resources – How Much Water Can I Save?* www.environment-agency.gov.uk.
[9]   New Mexico Office of the State Engineer. *A Water Conservation Guide for Commercial, Institutional and Industrial Users*. July 1999.
[10] OGBC Buying solutions. *Managed Services – Welcome to Watermark*. www.ogcbuyingsolutions.gov.uk/energy/watermark.

[11]   Integrated Pollution Prevention and Control (IPPC). *Reference Document on Best Available Techniques for Mineral Oil and Gas Refineries.* European Commission. Integrated Pollution and Control Bureau. Joint Research Center. Seville, Spain. 2003.

[12]   Muller K. and Sturm A. *Benchmarking Corporate Ecology of the Finance and Insurance Industry in Switzerland, Germany and Austria.* Ellipson AG. www.ellipson.com. August 2000.

[13]   Federal Energy Management Program. *Federal Water Use Indices.* http://www.eere.energy.gov/femp/technologies/water_useindices.cfm.

[14]   Environment Agency. *Water Resources: How Much Water Should We be Using?* www.environment-agency.gov.uk.

[15]   North Carolina Department of Environment and Natural Resources. *Water Efficiency Manual for Commercial, Industrial and Institutional Facilities.* p. 74, 1998.

Chapter 4

Measuring Flow and Consumption

To manage water it is essential to measure it. The primary uses of meters are to measure flow rates and consumption. Dataloggers capture the information and communicate this electronically. This chapter is devoted to a brief discussion of meters and dataloggers.

4.1 Flow Measurement

The ability to measure flow is critical to a water conservation programme. Selecting the right flowmeter for the task at hand will ensure that the results are accurate and repeatable. It will also provide value for money by minimising purchasing, installation and maintenance costs.

There are many technologies for measuring flow in closed conduits. This number is rising yearly. The advent of microelectronics in flowmeters has significantly improved their accuracy, ease of use and maintenance.

However, a highly accurate water meter is not required for all occasions. The best method of measurement is dependent on a number of factors:

- How accurate should the flow measurement be?
- Can the measurement be done manually? For instance, a flow cup such as in Figure 4.1 may suffice when measuring flow from taps. Or even a calibrated container and a stop watch could be used for measuring showers and basins.
- When were the meters calibrated? How old are the meters? Generally a turbine-type meter has a life expectancy of 10 years
- What is the expected flow to be measured? High velocities can wreck some types of meters.
- Are there any existing meters on-site that can be connected to a data-logger to be read remotely?

Figure 4.1   A photo of a flow cup

Courtesy of Con-Serv Pty Ltd.

- Are the readings required over a period of time such as 4–6 weeks or will a one-off measurement suffice?
- What is the pressure and temperature within the pipes?
- Are intrinsically safe meters and dataloggers required such as in oil refineries?
- Is the water ultra pure? Ultrasonic flowmeters are not suitable for such conditions. In dirty water systems mechanical meters may get blocked
- Is the meter easily accessible? Meters need to be accessible to be read and calibrated.

For measurement of water flows from taps the flow cup shown in Figure 4.1 will suffice. It is inexpensive, easy to use and the accuracy is generally acceptable. To measure water flow using a flow cup, simply place the cup under the tap, turn flow cup mechanism until the water does not overflow. At this point read the graduation on the side of the cup which can be read in litres per minute.

For water audits in commercial and industrial facilities to detect leaks and compute water balancing, it is essential to monitor the main water meter/s and install sub-meters in the main water-using areas. A brief description of the common water-meter types and their characteristics are given below.

In selecting a flowmeter, the desirable characteristics are the following:

- Measure flow with high accuracy over a wide range of operating conditions (e.g. low flow and high flow).
- Offer easy installation in any section of the piping without having to shut down the system.
- Be rugged so as to function reliably for many years without scheduled maintenance.

- Generate little or no pressure drop.
- Reasonable initial cost with low maintenance and operating costs.

Below a brief description of flowmeters suitable for water-flow measurement is given.

4.2 Types of FlowMeters

Meters can be classified into five general categories, based on the way they work. These are shown in Table 4.1

The focus of this chapter will be on positive displacement and velocity meters which are described below.

4.2.1 Positive Displacement Meters (volumetric)

Positive displacement also known as volumetric refers to the movement of the meter's flow-sensing element which displaces a specific volume for each cycle. Positive displacement meters include the following types: volumetric,

Table 4.1 Classification of meters

Type of meter	Mode of measurement	Examples
Positive displacement	A known volume of water is physically displaced by the measuring mechanism.	Volumetric. Generally for small-diameter pipes between 15 and 50 mm pipe diameter (5/8–2 in.)
Velocity	Use the rate of flow to measure volume. They convert velocity measurements to flow measurements.	Propeller, turbine, ultrasonic, electromagnetic and differential pressure.
Compound	Used to measure large variations in flow by having large meter and a small meter in the same unit.	A turbine meter on the main line and a small meter on a bypass line, with a valve to direct water to one or the other meter automatically. Generally installed when the meter size exceeds 100 mm in diameter (4 in.).
Proportional	A bypass meter measure only a small proportion of the total flow. A multiplier in the meter then multiplies and records the total flow through the whole unit.	Used in water towers and fire lines. Relatively accurate for large flows. Not suitable for low flows.
Open-channel	Used for flow measurement in open channels.	Weirs, flumes are two types used. Usually used for measuring wastewater flows.

Figure 4.2 V100 Piston-type volumetric meter

Courtesy of Elster Metering Pty Ltd.

nutating disk, oval gear, rotating vane, oscillating piston and reciprocating piston. The flow rate is calculated based on the number of times these compartments are filled and emptied and multiplying it by a constant established for volume per cycle. Positive displacement meters are well adapted to applications requiring high accuracy, low flow rates over widely divergent flow rates with relative low pressure head loss through the meter. These are available for pipe diameters 15 mm (5/8 in.) to 50 mm (2 in.). Figure 4.2 shows a popular volumetric meter.

Piston meters have a piston that rotates within a volumetric outer chamber as water flows through the meter. As the piston completes each revolution a known volume of water is displaced. By coupling the inner piston to a counter via transmission system to a register through a magnetic drive and gear assembly, the total volume of water can be recorded or transmitted through a probe to a datalogger. The nutating disk is another example of a positive displacement meter.

Positive displacement meters cannot be calibrated. Any undue clearances (due to wear and tear or from impure water sources) between the rotating elements and the outer chamber in which it rotates allow water to pass through without being registered.

4.2.2 Velocity Meters

Velocity meters operate on the principle that water passing through a known cross-sectional area with a measured velocity can be equated into a volume of flow as per equation (4.1). Velocity meters are suitable for high-flow applications.

$$Q = V \times A \qquad (4.1)$$

where

Q liquid flow through the pipe
V average velocity of the flow
A cross-sectional area of the pipe.

Velocity meters consist of

mechanical meters – turbine, propeller and multi-jet and,
non-mechanical – ultrasonic, electromagnetic and differential pressure types
   such as orifice meters.

Each type has its advantages and disadvantages when used in different operating conditions. For potable water applications generally mechanical meters are suitable. They are relatively inexpensive to purchase and install but require more maintenance since the moving parts are affected by abrasive substances. Non-mechanical meters are suitable for non-potable water which contains debris, require less maintenance but are more expensive.

4.2.2.1 Mechanical Meters

Turbine meters – A common type of mechanical meter is the turbine meter (Figure 4.3). Turbine meter operates by a turbine suspended axially to the direction of flow in a pipe. As the water passes through the pipe the turbine spins at a rate proportional to velocity of water flow. The blades rotation is detected by non-contacting sensor, and the volumetric flow rate is inferred from this measurement. The flow coefficient is calibrated in the laboratory and expressed in number of pulses/unit of volume. Applicability is in clean water systems. The meter is normally mounted with flanges to the existing pipe work. The output from the meter can be via a pulse from a reed switch

Figure 4.3   A photo of a turbine meter

sensor and/or conventional mechanical counter or register. Other types are propeller or paddle wheel meters. These are similar to turbine meters.

4.2.2.2 Non-Mechanical Meters

Electromagnetic – Electromagnetic meters require a conducting fluid such as water to measure flow rate. When a conductor moves across a magnetic field, a voltage is induced in the conductor, and the magnitude of the voltage is directly proportional to the velocity of the flowing water, which as per Equation (4.1) is proportional to the flow rate.

The magnetic flowmeter is a true volumetric device unaffected by fluid properties. The flowmeter is free of obstructions and therefore creates negligible pressure loss. The conductivity of water needs to be greater than $1\,\mu S/cm$. Magnetic flowmeters are popular in the water treatment, mining, pulp-and-paper and food-processing industries. Their accuracy is 0.2–1% and they can be installed in 2.5–2500 mm diameter pipes. The output can be analogue flow rates (L/min) and/or consumption similar to the register of a conventional mechanical meter.

Ultrasonic meters – There are two types of meters – transit-time method and Doppler method. The transit-time type calculates velocity based on the time difference for an impulse to pass between two sensors (transducers) located on opposite sides outside of the pipe. When a sound wave is transmitted between the transducers in the direction of flow, its velocity increases over that in still water. Similarly, its velocity decreases when transmitted against the flow. The water velocity is calculated as a function of the difference between the upstream and the downstream transmission times, the angle between the wave path and the water flow, and the speed of sound in quiescent water (which is a function of the density of water).

The Doppler-type meters measure the frequency of sound signals that are sent through the water flow from a transducer head. The signals are reflected back to the transducer head from moving particles and air bubbles in the water. Based on the assumption that the suspended particles are travelling at the same velocity as the water stream, the water velocity becomes a function of the reflected sound waves in accordance with the Doppler principle.

Transit time meters are used on clean liquids, while Doppler-type meters are generally used on slurries or liquids with entrained gas. Ultrasonic meters have a typical accuracy between 1 and 5% of flow rate. Their attraction is lack of moving parts, fast response; essentially zero head loss, and bidirectional measurement capability. The velocity of sound in water varies by about 0.2% for every 1° C change in temperature, so measurement must be compensated to minimise errors. Figure 4.4 shows an ultrasonic flowmeter.

Ultrasonic transit time meters (Figure 4.4) are also available as portable hand-held instruments with clamp-on transducers suitable for short-term investigative work where critical applications cannot be shut down to allow an in-line meter to be cut in. These can provide velocity, volumetric and totalled flow with built-in datalogger and rechargeable battery.

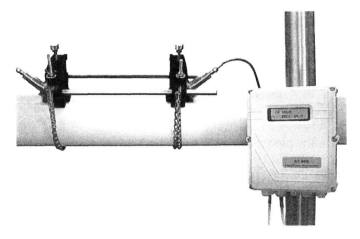

Figure 4.4   A photo of an Ultrasonic meter

Courtesy of GE Panametrics.

Other types of non-mechanical flowmeter types are orifice meters, rotameters, venturi tubes, pitot tubes and vortex shedding. Table 4.2 summarises the characteristics of flowmeters.

Table 4.2 Summary of velocity type flowmeters

	Most suitable application	Typical Accuracy	Pressure loss	Relative cost	Minimum straight run length of piping required upstream of meter in diameters
Mechanical (Turbine)	Clean liquid	0.1–1.0%	High	Medium	10–20
Magnetic	All types of liquids	0.2–1%	Low	High	5
Ultrasonic – Transit time	Clean and corrosive liquids. Some proprietary meters are capable of being used in dirty water such as raw sewage	2–5%	Low	Medium/ high	5–20
Ultrasonic – Doppler	Dirty liquids, corrosives and slurries	1–5%	Low	Medium/ high	5–20
Rotameter	Clean liquids	0.5–5%	Medium	Low	None
Orifice plate	Clean liquid	0.75%	High	Low	10–30
Vortex shedding	Clean liquid	1%	Medium	Medium	15–25

4.3   Selecting a Flowmeter

Given the range of meters, the questions to be asked when selecting a meter in no particular order of priority are

- What is the budget?
- Is it a permanent installation?
- Is the water clean or dirty?
- What is the pipe diameter?
- What is the desired accuracy?
- Will the flow rate be constant?
- What rangeability of measurement is required?
- Do pressure-loss limitations exist?
- Is an output signal required?
- Is it to be installed in an intrinsically safe environment?
- What is more critical: accuracy or repeatability?
- How much straight piping is available upstream or downstream of the flowmeter?
- Is excessive vibration likely to occur in the pipeline?
- Is the flow steady or pulsating?
- Will the pipe be constantly filled with water?
- What will be the orientation of the water meter be – horizontal or vertical?
- Is power available to the site or requires battery or solar power?
- Can the process be shut down to install the flowmeter?

4.4   Dataloggers

Most water meters provide only cumulative consumption readings and/or instantaneous flow rates. They do not provide time of use (TOU) consumption histories needed for the forensic analysis of system problems. To obtain such valuable TOU information, digital dataloggers are essential.

Dataloggers are low power devices typically powered by a long-life internal battery or external power sources such as solar panels. Dataloggers can be of the static or dynamic type. Static dataloggers need to have the data manually downloaded either in the field or on return to base. Dynamic loggers on the other hand have an inbuilt communications system that allows data to be automatically transmitted back to base. Static dataloggers are inexpensive and simple to use suitable for on-off logging.

Figure 4.5 shows a portable light-weight datalogger that can provide an output to a laptop computer. A solar-powered datalogger suitable for a remote site is shown in Figure 4.6.

Dataloggers have an inbuilt clock and microprocessor to time-stamp meter pulses counted over a user-definable interval (e.g. 5 minutes) and record data over long periods (e.g. 2–3 months but depending on the sampling rate).

Figure 4.5 A portable highly versatile datalogger

Courtesy of Hastings Dataloggers Pty Ltd.

Figure 4.6 Photo of a solar-powered datalogger

Courtesy of National Project Consultants Pty Ltd.

Dynamic loggers are best for long-term permanent monitoring and automatic meter reading. Transmission to base is through a telemetry system linked to a central data system or to a host website. Transmissions between dataloggers and computers can be automatically error-checked to ensure accuracy of data transfer.

The common telemetry systems are

• Public switched telephone network – The datalogger is connected to a modem and linked to the public telephone network. Retrieval is via a telephone and receiving modem. Low set-up costs and ongoing costs are limited to call costs. Whilst reliable they require good access to a landline.

- Radio networks – Radio networks offer the advantage that they can be used in locations without landline telephone networks. The types of radio networks used are spread spectrum and single channel.
- Mobile telephone networks – These are based on GSM (Global System for Mobile communications) common in urban areas. The data is transmitted using packet switches commonly known as GPRS (General Packet Radio Service). Portable and reliable but can be expensive to operate. Code Division Multiple Access commonly known as CDMA has a greater coverage and higher data transfer rates compared to GSM because they use multiple channels transmitted over multiple frequencies. They also have ongoing call costs and high power consumption.
- Satellite Telephone networks – These do not have the limitations of CDMA and GSM technologies and have better coverage. However, they suffer from higher initial and operating costs.

4.5   Chemical Methods of Flow Measurement

Under certain circumstances chemical concentration methods can be used to estimate the flow through pipes or open channels. This method is based on injecting a known amount of a readily soluble salt such as NaCl and then measuring the conductivity and time taken to reach a known distance. Fluorescent dyes can be used instead of NaCl in non-potable waters that are not highly coloured. Some water-treatment companies also market proprietary systems under various trade names.

4.6   Conclusion

In conclusion, without measurement there cannot be control. The meters need to be selected with due consideration given for type of use, cleanliness of the water, budget and degree of accuracy required. Dataloggers will enable the user to monitor the data on a continuous basis. For high water users it is essential to have permanent monitoring systems.

Chapter 5

Cooling Water Systems

5.1 Introduction

Heat is a by-product of human activity. Comfort cooling, machines and industrial processes all generate a large amount of heat. This heat must be dissipated continually – and flowing water is a good medium to remove heat – which is dissipated through evaporation.

Industries such as steel, aluminium, chemical and oil refining involve a number of heat-intensive processes. To control these processes and protect process-equipment from overheating, cooling water is used to transfer waste-heat from the process. Cooling water use – and associated evaporative losses – constitutes the highest consumptive use of water in most of these industries.

Water more than any other medium is used for cooling because it is

- readily available
- cheap
- has an ability to absorb large amounts of heat per square area (relative to air-cooled systems)
- can be discharged to the public utility sewers, creek or sea.

Therein lies the popularity of water in cooling systems.

Table 5.1 shows the percentage of overall water used in cooling systems in different industries. Power plants – by far – are the largest users of cooling water, as they require large amounts of cooling water to condense steam. For example, a 1000 megawatt (MW) power plant requires about 1500 m³/minute (400 000 US gal./minute) cooling water. This chapter is, however, devoted to industrial and commercial cooling towers – which are by far – the most numerous.

There are a number of water sources that can be used as cooling water such as potable water, river water, ground water, sea water and recycled water (tertiary treated sewage effluent) and other process streams. Salt water is abundantly available at coastal locations. The use of ground water for cooling purposes is now under increasing scrutiny.

Table 5.1   Percentage of water used in cooling systems

Industry	Water used in cooling systems as percentage of total
Office building	30%
Food manufacturing – refrigeration	10–30%
Cool rooms	80%
Oil refineries and chemical plants	60–85%
Steel mills	87%
Power plants	90–95%

5.2   Types of Cooling Systems

There are three types of cooling water systems. These are

1.   once through
2.   closed circuit
3.   open recirculating.

Once-through systems – They require large amounts of water – since the water only circulates once within the system. Once-through cooling systems are commonly found in thermal power plants (and other industrial plants) located in coastal areas which have access to sea water for cooling. These cooling systems due to their potential for environmental damage from thermal shocks and pollution are under increasing regulatory scrutiny and are being phased out in some countries.

Closed circuit systems – In closed circuit cooling systems the cooling water is completely confined within the systems pipes and heat exchangers. Reliable temperature control for industrial processes – such as the cooling of gas engines and compressors – is often achieved by closed recirculating cooling. Other applications for closed cooling systems are in chilled water and refrigeration systems. In such systems, water circulates in a closed cycle and is subjected to alternate cooling and heating – without air contact. Heat absorbed by the water is normally transferred by a water-to-water exchanger to the recirculating water of an open recirculating (or a once-through) cooling system. Closed-loop systems are not designed to lose water except for leaks.

Open recirculating systems – Open recirculating cooling water systems are the focus of this chapter as they have the greatest potential to save water.

The attributes, advantages and disadvantages of each system are shown in Table 5.2.

Table 5.2 A comparison of three cooling systems

System	Advantages	Disadvantages	Capacity of industrial process
Once through direct	• Low capital cost • Low operating cost • Low temperature sink	• Obtaining water intake and discharge permits may be difficult • Imparts thermal shock • Biological fouling from intake water • Pollution of waterways • Expensive heat exchange metallurgy costs for sea water or brackish water	<0.01– >2000 MW (0.034–6823 MMBtu/h*)
Open recircu-lating	• Low make-up water rate • Low discharge water rate • Good chemical control	• High capital cost • Power for fans and pumps • High chemical treatment costs • Risk of Legionella	<0.01– >2000 MW (0.03–6823 MMBtu/h)
Closed recircu-lating	• Minimum make-up water rate • High quality make-up • No biological fouling • No risk of *Legionella* • Minimal chemical requirements • Carbon steel acceptable for most components	• Expense of secondary cooling system • Metallurgy of water – water heat exchanger set by final heat sink • Heat load limitations, including temperature approach to final heat sink • Additional pumps and piping required	0.2–10 MW (0.7–34 MMBtu/h)

* MMBtu/h-million Btu/h

5.2.1 Open Recirculating Cooling Water Systems

Cooling towers, due to their apparent simplicity, are often taken for granted and in commercial buildings they are often located on top of the building – rendering them even less visible. So the old adage out of sight – out of mind often applies to cooling towers.

Table 5.1 shows that cooling water systems can use a large percentage of the total water intake and therefore need to be considered in any water conservation programme.

The primary purpose of cooling water systems is to cool process fluids so the plant can operate at maximum efficiency. Lack of attention to cooling systems can result in

- plant shut-downs due to scaling and the corrosion of critical heat exchangers
- increased risk of Legionellosis
- large quantities of water wastage
- increased water costs, wastewater costs and chemical treatment costs.

The Melbourne Aquarium case study serves as a grim reminder of the risks posed by cooling towers.

Fouling and poor maintenance of cooling tower fill can lead to a deterioration in thermal performance of the cooling tower and consequently an increase in cold water temperatures – leading to scaling potential of heat exchangers.

Case Study: Melbourne Aquarium, Australia [1]

The Melbourne Aquarium was opened in early 2000 and became a very popular tourist destination.

In April 2000, Legionnaire's disease was identified. It killed three people and seven were hospitalised in a critical condition. Ninety-three (93) people contracted the disease. The source of the pathogen were the cooling towers. Despite replacement of the cooling towers with air cooled chillers, attendance dropped by almost 50%. The court awarded the victims A$450 000 to be shared amongst them. Future litigation by the affected parties is a possibility.

5.2.1.1   Recirculating Cooling Water Systems – Operational Principles
The principle of a recirculating cooling tower, whether it is in a power plant or in an air-conditioning system is the same.

Figure 5.1 shows the basic equipment in a typical recirculating cooling system.

Cold cooling water at temperature T_{cw} is recirculated by the recirculating water pump. The water picks up heat from the heat exchangers (condensers). The hot water at temperature T_{hw} enters the cooling tower at the top. The water is sprayed through nozzles to the plastic or timber splash fill to disperse

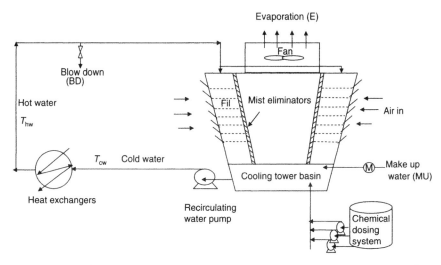

Figure 5.1 A schematic of a typical recirculating cooling system induced draught cross-flow cooling tower

the water into small droplets. The droplets exchange the heat with the incoming air flow that is sucked into the cooling tower by the induced draft fan.

The basis of cooling tower operation is sensible heat transfer and evaporative cooling. Sensible heat transfer takes place between the air stream and the hot water droplets. It accounts for only 25% of cooling. The bulk of the cooling (75%) takes place due to evaporation of water droplets when it comes into contact with an air stream. Water has a very high latent heat of evaporation (equal to 2431 kJ/kg or 1000 Btu/lb) compared to sensible heat transfer (4182 J/kg at 50° C) which is 582 times more. So, for every kg or pound of water that is evaporated, it removes 2431 kJ/kg (1000 Btu/lb) of heat from the rest of the water body. This results in cooling the rest of the water body. The evaporated water vapour leaves the cooling tower as hot moist vapour. Under normal operating conditions this approximates to 1.2% for each 5.5° C (10° F) temperature rise $(T_{hw} - T_{cw})$.

Thus, cooling water requirements are dictated by the amount of heat that needs to be removed from the hot cooling water.

Since only pure water and volatile impurities evaporate, all the non-volatile impurities (such as dissolved solids including calcium and magnesium hardness, chlorides, iron and suspended solids) concentrate in the cooling water. A constant bleed (BD) is maintained by regulating the blowdown valve to maintain a desired dissolved solids level, to minimise the impact of these impurities on the cooling water system. A freshwater makeup line maintains the water level in the cooling tower.

A small amount of water also is lost as fine droplets (entrained) in the air stream. This is called "windage" or commonly known as "drift" (DR). Drift is defined as the water lost from the tower as liquid droplets entrained in the exhaust air. It is independent of water lost by evaporation. Unlike evaporation, drift contains dissolved solids that are in the cooling water. Drift eliminators reduce drift loss by removing entrained water from the discharge air by causing sudden changes in direction of air flow. The resulting centrifugal force separates the drops of water from the air, depositing them on the eliminator surface from which they flow back into the tower. Drift is expressed as a percentage of the recirculation rate and typically ranges between 0.1 and 0.002% of the recirculation rate. The rate of drift loss depends on the age and design of the tower.

Splashing from the cooling tower is another source of water loss. This is generally due to poor maintenance of fill, or strong wind blowing through the tower.

Make-up water (MU) is added to replenish the water lost from evaporation, blowdown, drift, splash and other areas.

5.2.1.2   Recirculating Cooling Water Systems – Basic Concepts
In cooling tower design the variables of importance are

- The heat load imposed on the tower (Q)
- Cooling water flowrate (RR)
- Cooling range ($T_{hw} - T_{cw}$)
- Ambient wet bulb temperature (T_{wb})
- The approach temperature defined as the difference between the cold water temperature and the wet bulb temperature of air ($T_{cw} - T_{wb}$)
- Evaporation rate (ER)
- Make-up water (MU) flow rate
- Concentration ratio (CR)
- Blowdown (BD).

These concepts are explained below.

Calculating the heat load Q.   Cooling water requirements are dictated by the heat transfer load. The heat transfer load is the sum of all heat loads from all exchangers and can be calculated from Equation (5.1):

$$Q = \sum UA(\text{delta } T_m) \tag{5.1}$$

where

Q – Heat transfer, kW (Btu/h)
U – Heat transfer coefficient, $W/(m^2)(°C)$ or $[Btu/(h)(ft^2)(°F)]$
A – Heat transfer surface area, $m^2 (ft^2)$
Delta T_m – Mean temperature difference between fluids, °C (°F).

Cooling water requirement. Once the heat transfer requirements for each heat exchanger has been calculated, the volume of cooling water needed to remove the heat is determined using Equation (5.2 or 5.3):

$$CW = Q/[4.2(T_{hw} - T_{cw})] \qquad L/s \tag{5.2}$$

For US units,

$$CW = Q/500(T_{hw} - T_{cw}) \qquad US\ gal./min \tag{5.3}$$

[The unit conversion factor of 500 is derived from 8.33 lbs/gal. × 60 min /h × 1 Btu/(lb)(°F)].

The overall cooling water requirement is the sum of CW plus any other equipment that uses cooling water.

Cooling range $(T_{hw} - T_{cw})$. The cooling range $(T_{hw} - T_{cw})$ is independent of ambient temperatures and is only a function of heat load. For air-conditioning systems with electric compressors this is generally around 5.5° C (10° F). In manufacturing plant cooling systems this typically can be in the range of 5° C–30° C (9° F–54° F).

Ambient Wet Bulb Temperature T_{wb}. The ambient wet bulb temperature is the temperature achieved when a thermometer covered with a moist porous material is placed in an unsaturated flowing stream of moist humid air. The evaporation from the porous material causes a heat loss from the bulb, which causes a drop in the bulb temperature and hence in the thermometer reading.

The value that is equalled to, or exceeded by, 5% of the time – during the four hottest months – is generally the ambient wet bulb temperature chosen for cooling tower design.

Approach temperature $(T_{cw} - T_{wb})$. The difference between the ambient wet bulb temperature T_{wb} and the cold water temperature T_{cw} has a pronounced effect on the cooling tower size and cost. It is impossible for a cooling tower to cool water below the wet bulb temperature T_{wb}. Therefore designers try to minimise the difference.

However, there is a trade-off between performance and capital cost. For an example, by reducing the approach temperature from 8.3° C to 4.4° C the tower requirements increases by 50% and the fan operating cost by 65%. Thus various approach temperatures need to be evaluated to determine the minimum cost to the total system, including the heat exchanger surface areas, piping network, pumps and cooling tower costs.

Typical approach temperatures range from 2.8° C to 11° C (5° F– 20° F). Once the approach temperature has been selected it is important to operate as close as possible to this to minimise operating expenditure on energy.

In air-conditioning systems an increase of T_{cw} by 1°C increases the energy required by 3.5% to 4%. This is because as the cold water temperature increases, less heat is removed from the refrigerant gas and more energy is therefore required to compress the gas.

The relationship between approach temperature and cooling range is shown in Figure 5.2 below.

Evaporation rate.   Once the cooling water circulation rate and the cooling range is known it allows one to calculate the expected evaporation rate (ER) from the cooling tower as given in Equations (5.4 and 5.5):

$$ER = \frac{\text{Thermal load}}{\text{Latent heat of evaporation}}$$

$$= 0.8 \times RR \times (T_{hw} - T_{cw})°C \times 4.2/2431 \text{ kg/s} \qquad (5.4)$$

where

RR – pump recirculation rate L/s
0.8 – evaporation factor (typically ranges from 0.75–0.85).
4.2 – specific heat of water (kJ/kg °C)
2431 – latent heat of evaporation of water, kJ/(kg)(°C).

In US units,

$$ER = 0.8 \times RR \times (T_{hw} - T_{cw})/1000 \qquad \text{US gal./min} \qquad (5.5)$$

where

RR is in US gal./min
1000 is the latent heat of evaporation of water, Btu/(lb)(°F).

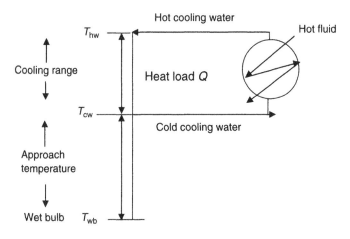

Figure 5.2   Schematic of thermal variables in a cooling tower

Make-up water required.   Make-up water (MU) is added to replace evaporation (ER), blowdown (BD) and drift loss (DR) and system losses (L) as given in Equation (5.6).

By material balance,

$$MU = ER + BD + DR + L \qquad (5.6)$$

Studies conducted by Sydney Water [2] on cooling towers in commercial buildings showed that on average 88% of the water is evaporated, bleed accounted for 5%, and drift and splash accounted for 7% of the make-up water.

The MU is normally measured using flowmeters. In the absence of a flowmeter, MU can be estimated using Equation (5.7).

$$MU = ER \times (CR/CR - 1) \qquad (5.7)$$

where CR is the cycles of concentration.

Cycles of concentration.   Cycles of concentration (CR) measure the number of times the water circulates within the cooling system before lost to blowdown. It is the ratio of an ion in the cooling water (C_{cw})/ to that of the make-up water (C_{mu}).

$$CR = C_{cw}/C_{mu} \qquad (5.8)$$

Typically the ions considered are magnesium or silica. For convenience, total dissolved solids (TDS), conductivity or chlorides (when the chlorides contribution from chlorination is insignificant) are also used.

The objective is to maximise CR so that water loss through blowdown is minimised. The maximum CR that can be maintained in a cooling water system is subject to the salinity of make-up water quality and other considerations such as system metallurgy.

Example
Calculate the cycles of concentration if the cooling water chloride concentration is 150 mg/L and make-up water chloride is 30 mg/L using Equation (5.8)

$$CR = C_{cw}/C_{mu} = 150/30 = 5 \text{ cycles.}$$

Blowdown.   Voluntary blowdown (BD) is the water that is purposely removed from the system to maintain concentration of solids within a desirable range. It can be measured using flowmeters or (where impractical) can be determined as follows:

By mass balance, the input solids = output solids.

$$MU \times C_{mu} = BD \times C_{cw} \qquad\qquad BD = MU \times C_{mu}/C_{cw}$$

Therefore substituting Equation (5.8)

$$BD = MU/CR \qquad (5.9)$$

$$CR = MU/BD = C_{cw}/C_{mu} \qquad (5.10)$$

The objective when operating cooling water systems is to maximise CR, to reduce water and chemical use and minimise BD.

System losses.   Water in cooling systems is lost through pumps, gland seals, washing of floor areas and splash from cooling towers. If there is a mismatch between the cycles of concentration calculated as MU/BD and when calculated as C_{cw}/C_{mu}, then it is a clear indication that leaks are occurring. The calculation concentration (or CR) as makeup divided by BD could be artificially high in systems that are leaking. Therefore, a better way to calculate the CR is using dissolved solids C_{cw}/C_{mu}.

From the overall water balance we get,

$$BD = MU/CR = ER/(CR - 1) - DR - L \qquad (5.11)$$

5.3   Types of Cooling Towers

Cooling towers are classified as natural draught and mechanical draught. Natural draught cooling towers found in electric power plants are hyperbolic with water flows as much $136\,000\,m^3/hr$ (600 000 US gpm). These are not the focus in this chapter.

Mechanical draught cooling towers are most commonly found in buildings and process industries. Fans are used to provide air flow of a known volume through the tower therefore improving thermal performance and stability. Cooling towers are commonly characterised by the location of the fan in the tower and by the direction of airflow. There are four different types of towers. These can be classified by location of fans and air-flow movement.

By location of fans:

- Induced draught
  - the fans are located at the top of the tower
  - the fans suck the air.
- Forced draught
  - the fans blow air
  - the fans are located at the inlet side of the tower.

By movement of airflow:

- Cross-flow towers
  - the air is admitted from the sides of the tower.

- Counter-flow towers
  - the air is admitted at the bottom of the tower and travels vertically upwards through the fill against the direction of the water flow.

Cooling towers are a combination of location of the fan and air flow. Thus they can be either induced draft counter-cross-flow or forced draft counter-cross-flow.

The selection of the most appropriate cooling tower is a function of

- required capacity
- cleanliness of the water
- location of the tower
- space availability
- operating efficiencies
- capital costs of the different types of towers.

In explaining the basics of cooling towers, only induced draft towers are considered in detail.

5.3.1 *Induced Draught Cross-flow Cooling Towers*

In cross-flow towers, the air enters the tower through the louvers that extend from the basin curb to the distribution deck on the sides of the tower – and travels horizontally – through the tower fill and drift eliminators. It then flows vertically upwards through the fan stack. The water flows from the open distribution system near the top of the tower – on the sides vertically down through the fill – and into the basin. Figure 5.1 is a schematic of an induced draught cross-flow cooling tower. Induced draught cross-flow towers are still preferred in applications where the circulating water can become contaminated with debris, for example in steel mill applications.

5.3.2 *Induced Draught Counter-flow Cooling Towers*

Counter-flow towers owe their popularity to the highly efficient film fill. Film fill is manufactured from formed sheet polyvinylchloride into a "honey comb" appearance with a large wetted surface area and with a number of various sized openings available. As a rule of thumb, 1 m of film type fill can be equivalent to 10 m of the traditional splash type fill, thus increasing the air/water contact area. This makes the counter-flow towers space efficient. Thermal efficiency is also greater because the counter-flow tower's average driving force for heat and mass transfer is greater than cross-flow since the coldest water contacts the coldest air and the warmest water contacts the warmest air [3].

Counter-flow cooling tower fill is less exposed to sunlight therefore reducing algae growth. However, this feature also makes them harder to service

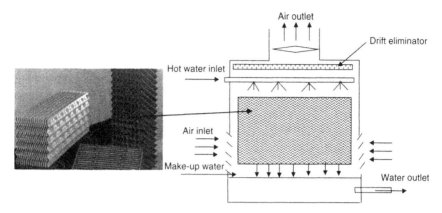

Figure 5.3   Schematic of induced draught counter-flow cooling tower with photo of fill

Photograph courtesy of Baltimore Aircoil.

and more prone to a deposit build up. Figure 5.3 shows induced draught cooling tower with a photo of cooling tower fill.

5.3.3   Forced Draught Wet Cooling Towers

The difference these ones have is that the cooling tower fans are located at the base of the cooling tower thus forcing the air through the tower. Thus they are easy to maintain. Normally these are of counter-flow design.

5.3.4   Evaporative Condensers

These are normally found in the refrigeration industry, where the refrigerant is cooled directly by the water circulating in the tower (known as the evaporative condenser). These are also known as mechanical draught wet closed cooling systems. They are characterised by having a very shallow basin and a small cooling water circuit. The hot refrigerant gas (normally ammonia in food processing plants) is cooled directly by the condenser water.

5.4   Water Conservation Opportunities

How can water conservation opportunities be maximised in cooling towers?
    The key areas to maximise opportunities are by

- reducing involuntary water loss
- reducing voluntary water loss
- improving operating practices

- substituting alternative water sources to potable water
- converting to alternative cooling systems (Chapter 6).

These methods are described below.

5.4.1 Reducing Involuntary Water Loss

Involuntary water loss can be minimised by

- minimising overflow of water from cooling tower basins
- adjusting incorrectly set piping configuration
- arresting leaks in tower basin joints, pump seals
- reducing filter backwash water flow rates
- reducing excessive drift
- reducing water splash.

It is worth spending some time on each of these areas.

5.4.1.1 Minimising Overflow of Water from Cooling Tower Basins
Incorrectly set or defective ball float valves are frequently the cause of cooling tower basin overflows. As Figure 5.4 shows, when the float valve is set too high, the water overflows through the overflow pipe. They require periodic inspections.

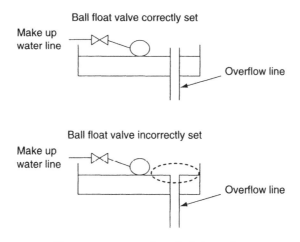

Figure 5.4 Incorrectly set ball float valves

Adapted from Sydney Water, *Best Practice Guidelines for Commercial Cooling Towers*, 2004.

5.4.1.2   Incorrect Piping Configuration

Incorrect piping configuration is quite common in air-conditioning systems. If the condenser water pipes run above the height of the tower spray heads, when the pump shuts down, water could flood back into the tower.

5.4.1.3   Leakage from Pipes, Joints and Pump Glands

With the passage of time, tower basin joints and other joints frequently leak. Leaks can be minimised through a well-managed maintenance programme. Pump gland leaks can be addressed in a timely manner by repacking them.

5.4.1.4   Drift Loss

All cooling towers lose water due to drift. Excessive drift results in water and chemical losses making it harder to control voluntary blowdown. It is also a nuisance when excessive drift leads to spotting of cars, buildings and windows. If the cooling tower drift eliminators have not been inspected recently, organise for a specialist to assess the performance.

5.4.1.5   Splash

Constant wetness around the cooling tower is an indication of splash. This may be due to high winds or a design flaw. Install anti-splash louvers to minimise splash.

5.4.2   Reducing Voluntary Water Loss

As mentioned earlier, a portion of the water is required to be blown down to minimise excessive concentrations of impurities in the water depositing on hot heat transfer surfaces, reducing thermal performance of the heat exchanger, causing corrosion or leading to fouling of tower internals.

The amount to be blown down is dictated by the quality of the make-up water. More often than not operators tend to blowdown excessively. As Figure 5.5 demonstrates, sometimes there is very little blowdown due to blocked blowdown valves.

The reasons for excessive blowdown are

- incorrect setting of conductivity controllers
- the water-treatment supplier not managing the cooling water-treatment programme (in the interests of the customer)
- poor operation and maintenance practices.

There are a number of ways of reducing voluntary water loss and these are described below:

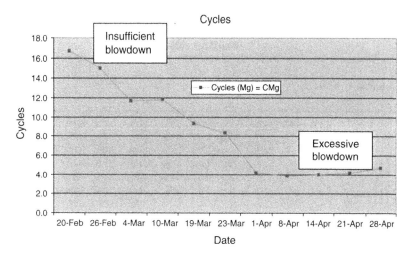

Figure 5.5   Cooling tower cycles in a shopping mall

5.4.2.1   Increasing Cycles of Concentration

Significant water savings can be achieved if the cycles are less than 5 in typical cooling water applications. The percentage of cooling water consumption that can be conserved by increasing the cycles of concentration is given in Equation (5.12).

$$\text{Per cent conserved} = CR_n - CR_i/CR_i(CR_n - 1) \times 100\% \qquad (5.12)$$

where

CR_i Cycles of concentration before increasing cycles
CR_n Cycles of concentration after increasing cycles.

Using Table 5.3 the percentage of make-up water saved can easily be calculated.

For example, by using Table 5.3 the percent make-up water that can be saved by increasing cycles from 3 to 10 is equal to 26% for any make-up water volume.

If we assume the following costs, then

Cost of water	$1.20/m^3
Cost of sewer usage charge	$1.20/m^3
Cost of chemical treatment, electricity	$6.0/m^3
Total cost of cooling water	$8.40/m^3
Initial make-up water flow rate	20 m^3/d
System operates for	300 days/yr 24 hrs/day

Table 5.3 Percent of make-up water saved

Initial concentration (CR_i)	New Concentration (CR_n)										
	2	2.5	3	3.5	4	5	6	7	8	9	10
1.5	33%	44%	50%	53%	56%	58%	60%	61%	62%	62.5%	63%
2	–	17%	25%	30%	33%	38%	40%	42%	43%	43.8%	44%
2.5	–	–	10%	16%	20%	25%	28%	30%	31%	32.5%	33%
3	–	–	–	7%	11%	17%	20%	22%	24%	25%	26%
3.5	–	–	–	–	5%	11%	14%	17%	18%	20%	21%
4	–	–	–	–	–	6%	10%	13%	14%	16%	17%
5	–	–	–	–	–	–	4%	7%	9%	10%	11%
6	–	–	–	–	–	–	–	3%	5%	6%	7%
7	–	–	–	–	–	–	–	–	2%	4%	5%

Adapted from North Carolina Department of Environment and Natural Resources. *Water Efficiency Manual.* August 1998.

Annual water savings
Water savings increasing from 3 to 10 cycles $5.2\,m^3/d$
Annual water savings $300 \times = 1560\,m^3$
Annual Cost savings $\$8.40 \times 1560 = \$13\,104$

Annual water and chemical treatment charges at 3 cycles equals $50 400.

The result is a slashing of the water and chemical treatment bill by 26% with no extra capital or operating expenditure.

5.4.2.2   Install Flowmeters on Make-up and Blowdown Lines and Conductivity Meters in Blowdown Lines

This will allow the operator to closely monitor the volume of water being used and verify that the system is operating at optimum cycles of concentration. Ideally all three signals need to be fed to a central data monitoring point so that the operators can remotely check the performance of the system.

5.4.2.3   Operate Blowdown in Continuous Mode

Cooling towers that have automatic conductivity controls are typically bled when the conductivity reaches a specified conductivity value releasing a large quantity of water. Blowdown control by this method leads to large fluctuations in the conductivity, which wastes water. A better method is to operate on a continuous bleed mode by keeping the conductivity close to the upper control limit. Statistical process-control software (supplied by reputed water treatment vendors) allows this to be achieved with greater accuracy. The upper and lower control limits are divided into one, two and

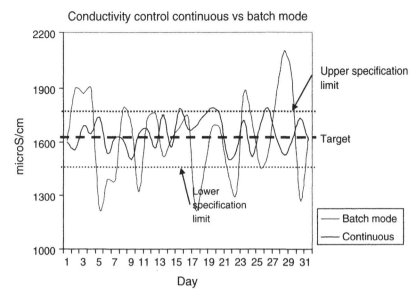

Figure 5.6   Continuous conductivity control vs batch mode

three standard deviations. The objective is to control within one standard deviation.

Figure 5.6 shows the difference between these two modes of control.

5.4.2.4   Install Sidestream Filtration

Cooling towers are good scrubbers of air-borne particulate matter, especially in dusty environments such as in steel mills. Suspended solids contribute to fouling of cooling system, plugging of cooling tower nozzles resulting in irregular distribution of cooling water in cooling towers, and in the tower sump, suspended solids accumulate and if not cleaned regularly create an environment for corrosive anaerobic bacteria to thrive.

In the cooling system, suspended solids contribute to clogged spray nozzles, erosion of piping, pumps and heat exchangers resulting in unscheduled plant shut downs.

The usual response from plant operators is to increase blowdown rates or to add chemical dispersants. Both these methods do not address the root cause. Blowdown wastes water and chemicals. The use of additional chemicals adds to chemical-treatment costs.

A smarter solution is to continuously remove a percentage of the solids loading using a sidestream filter.

Granular media filtration, bag and cartridge filters, hydrocyclones and self-cleaning screen filters are used for sidestream filtration. Typically, 1–20% of the recirculation flow is passed through a sidestream filter.

Modern automatic self-cleaning filters such as screen filters (as shown in Figure 5.7) can reduce the particle size down to 10 μm (if the particles are

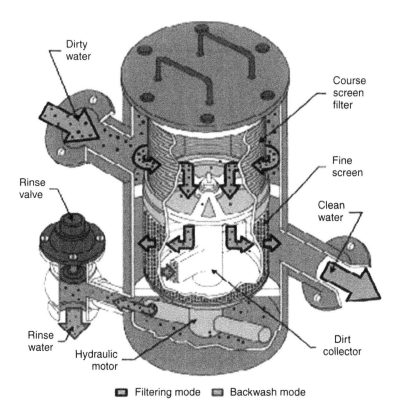

Figure 5.7   Schematic of a self-cleaning automatic filter

Courtesy of Automatic Filters Inc.

less than 25 μm it is considered an excellent outcome). In screen filters, the cooling water passes through a wedge-wire, woven-wire or perforated cylindrical elements with solid particles retained on the screen surface. They remove both organic and inorganic particles. The filters work on the principle of differential pressure. When the pressure differential which is normally 7–14 kPa (1–2 psi) is exceeded, the filter initiates the automatic cleaning cycle for about 40 seconds. The total volume of water used can be as low as 1% or less, which is significantly less than granular media filters use for backwash. Figure 5.8 shows how one of the ways an automatic self-cleaning screen filters is installed.

5.4.3   *Improving Operating Practices*

5.4.3.1   Shut Off the Unit When not in Operation
When the building is unoccupied or when the plant is not operating shut down the cooling water system.

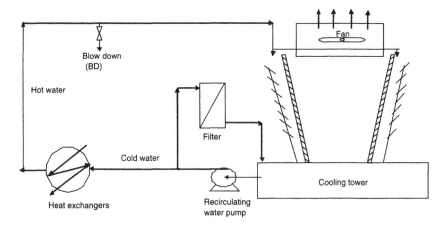

Figure 5.8   Schematic of a simple side-stream filtration system

5.4.3.2   Minimise Process Leaks to the Cooling System

Minimise process leaks to the cooling water system, such as glycol from the closed-loop system or hydrocarbons. Hydrocarbons increase microbial growth in the cooling water.

5.5   Alternative Water Sources

As shown in Table 5.4, there are a range of alternative water sources that can replace potable water. Cooling towers do not require water with the highest purity and therefore the alternative sources are generally cheaper than potable water. However this needs to be balanced again several criteria such as

- Health and safety requirements
- Impact on the cooling system performance
- Impact on heat exchanger metallurgy – corrosion
- Potential for fouling of heat exchangers
- Ease of control – cooling water chemistry
- Increased pumping requirements
- Security of supply – potential for supply disruption
- Increased chemical treatment costs.

5.5.1   Maximum Allowable Concentrations

The maximum allowable concentrations arising from the use of alternative water sources in a cooling tower are dictated by their potential to cause scaling, corrosion and fouling in cooling water systems. For each system

Table 5.4 Comparison of alternative water sources

Source	Positives	Negatives
River water	A secure source of supply	High in suspended solids, turbidity and organics. May also contain NH_3 and aerobic and anaerobic microbiological organisms resulting in reduced heat transfer and increased corrosion.
Ground Water	A secure source of supply	May contain high levels of hardness, TDS, Fe and Mn, Silica and alkalinity. Refer to Chapter 2 for more details.
Reverse osmosis reject water (brine) and demineraliser regeneration water	Good-quality water if the TDS is not too high	Can increase cooling water TDS levels as well increase the scaling and corrosion potential of water
Boiler blowdown	Boiler water frequently has a lower conductivity than city water	Hot. Contains polymers and phosphates that may interfere with cooling water chemistry. Increase phosphate loadings and is a nutrient for micro-organisms.
Steam condensate	Similar to boiler blowdown	Hot. Can contain ammonia and amines leading to increased microbial growth and corrosion of copper metallurgy.
Air-conditioning condensate (from air handling units)	A source of cool water	Variable source of supply. Condensate may only be available during the summer months.
Recycled sewage effluent from secondary and tertiary treatment plants.	PO_4 can reduce chemical treatment costs	Micro-organisms, NH_3, PO_4, biochemical oxygen demand (BOD), TDS, PO_4 and Cl. Refer to Chapter 2. PO_4 will contribute to Ca phosphate scale. If high levels of suspended solids are present these can settle out in low flow areas of the cooling system. Chemical treatment costs will be higher than that for potable water.
Rain and storm water	Suitable for buildings with large roof areas	Storm water may contain oil and other organics. Variable supply source.
Sea Water	Suitable for new plants	Cycles will have to be controlled to 1.5–2 cycles. Requires special towers and titanium or other exotic metallurgy.

Table 5.5 Typical concentrations in cooling systems

Contaminant	Source	Typical cooling system level mg/L	Potential problem
Aluminium	River water	0	Fouling
Amine	Steam Condensate	0	Microbiological growth and fouling. Corrosion of copper alloys
Ammonia	Steam condensate, recycled sewage effluent	0	Microbiological growth and fouling Corrosion of copper alloys
BOD	Various	Not usually measured	Microbiological growth and fouling
Ca as $CaCO_3$	Various	100–1000	Scaling
Chlorides	Sea water	<1500	Corrosion
Hydrocarbons	Oily water. Storm water	0	Fouling, microbiological growth
Iron	Ground water	0–3	Fouling
Manganese	Ground water	0	Fouling and corrosion
Phosphate	Treated sewage effluent, boiler blowdown	0–25	Calcium PO_4 scale
Suspended solids	Various	20–150	Fouling
TDS	All sources	<2000	Corrosion, scaling

these need to be individually evaluated. Using generic guidelines without consideration of other system conditions may expose the system to risk of system failure. Therefore no maximum values are given. Table 5.5 shows typical concentrations.

Case study: Petrochemical Plant, China

A petrochemical plant in China using 50–70% of river water as make-up consistently had high levels of nitrifying bacteria, iron-oxidising bacteria, *E. Coli*, Sulphate-reducing bacteria, fungi and algae in addition to high levels of organics, suspended solids, Ammonia. The river water composition was as follows:

pH	N/A	7.5
Ca as $CaCO_3$	mg/L	93
M – Alkalinity as $CaCO_3$	mg/L	160
Conductivity	μmhos/cm	880
Ammonia as NH_3	mg/L	32
Organics as TOC	mg/L	30
Chlorides as Cl	mg/L	170
Copper	mg/L	0.05
Iron as Fe	mg/L	2.0
Free chlorine	mg/L	0.2

Case Study: Duke University installs water reuse system [4]

Duke University, Orange County, USA has a chemistry building with an area of $12\,913\,m^2$ ($139\,000\,ft^2$). The facilities maintenance staff installed systems to collect and pump condensed water from the building cooling systems to existing cooling towers. The investment was only two sump pumps and piping. Condensate water replaced evaporated water in the cooling tower system that was previously made up with potable water. As a result, Duke University saved in excess of $7570\,m^3$ (2 million US gal.) in the first year of operation.

5.6 Cooling Water Treatment for Recirculating Water Systems

The objectives of cooling water treatment are to maintain system performance and minimise occupational health and safety hazards.

The major areas of concern are

- scaling
- fouling
- corrosion
- microbial growth.

All of these phenomena are interrelated and affect system performance. Contributing factors such as process leaks, high temperature, suspended solids, low flow, ammonia and organics as shown in the Figure 5.9 will exacerbate one or several phenomena. Therefore all of these factors need to be considered in a comprehensive cooling water-treatment programme.

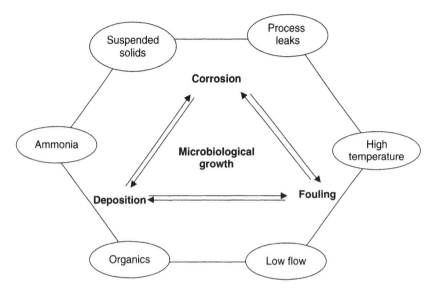

Figure 5.9   Critical factors affecting system performance

5.6.1   *Scaling*

The primary cause of scale deposition is due to supersaturation. The deposit forming salts exceed their solubility limits at a given temperature and deposition occurs. Common scale-forming salts are calcium carbonate (calcite), calcium phosphate, calcium sulphate and silicates. All of these exhibit inverse solubility with temperature. That is when the temperature increases the solubility decreases. Scaling reduces the performance of the heat exchanger; since the thermal conductivity of these scalants is significantly lower than that of steel. For example, the thermal conductivity of calcium carbonate is 25 times less than that of steel. Figure 5.10 shows the brick-like crystalline structure of calcium carbonate crystals and their large surface area allowing them to bind to each other.

As Figure 5.10 shows, scaling results in some of the heat exchangers becoming partially or totally blocked. As a result the heat transfer surface reduces resulting in the hot process stream (in the case of air-conditioning systems the refrigerant) leaving the exchanger hotter than specified. Moreover, scaling also results in the need to use extra pumps to pump water against the greater resistance of the blocked tubes.

The secondary causes of scale formation are

- Temperature
- Changes in water chemistry (Alkalinity, pH, ion concentrations)
- Pressure changes
- Flow rate changes
- Surface geometry.

Figure 5.10   Calcium carbonate crystals

Courtesy of New Logic Research Inc.

When these conditions are met the solubility of the sparingly dissolved substances are exceeded. These concepts are discussed in Chapter 2.

Calcium carbonate is by far the most common scale found in cooling water systems. As reaction 5.13 shows, it is the result of the breakdown of calcium bicarbonate. Therefore the degree of scaling is dependent on the calcium hardness and bicarbonate alkalinity in the cooling water.

$$Ca(HCO_3)_2 \longrightarrow CaCO_3 + H_2O + CO_2 \qquad (5.13)$$

Figure 5.11 shows $CaCO_3$ scaling in an air-conditioning condenser.

Figure 5.11 $CaCO_3$ scaling in condenser tubes

Courtesy of Absolute Water Treatment Pty Ltd.

5.6.1.1 Scaling Indices

There are a number of models to predict the scaling potential of $CaCO_3$ in water. They are

- Langelier Scaling Index (LSI)
- Ryznar Scaling Index (RSI)
- Practical Scaling Index (PSI).

The LSI is the oldest and is the most common. Dr John Ryznar and Paul Puckorius developed the RSI and PSI indices respectively and are variations of LSI.

The LSI is a model to predict the degree of saturation of water with respect to $CaCO_3$. The LSI approaches the concept of saturation using pH as the main variable. The LSI can be interpreted as the pH change required to bring water to equilibrium (LSI = 0).

The LSI is calculated using the following formula:

$$LSI = pH - pH_s \tag{5.14}$$

$$pH_s = pK_2 - pK_{spCaCO3} + pCa + pAlk \tag{5.15}$$

where

pH_s – pH at which $CaCO_3$ is at saturation
pK_2 – second dissociation constant for carbonic acid (from HCO_3^- to CO_3^{2-})
$pK_{spCaCO3}$ – Solubility constant for $CaCO_3$
pCa – minus of the logarithm of the molal concentration of Ca^{2+} in moles per 1000 g of water.
$pAlk$ – minus of the logarithm of the molal concentration of M Alkalinity in moles per 1000 g of water.
pH – Measured pH of the solution
M Alkalinity – Molal concentration of HCO_3^- to CO_3^{2-} and OH^-
Molal – gram moles of ion per 1000 g of water.

The rules for interpretation of LSI are given in Table 5.6.

For an example, water with an LSI of 2.5, LSI is 2.5 pH units above saturation. Reducing the pH by 2.5 units will bring the water into equilibrium.

Table 5.6 Rules for interpreting LSI

LSI	Saturation pH	Comments
+	pH of water > pH_s of $CaCO_3$	$CaCO_3$ precipitation likely. Scale can form.
−	pH of water < pH_s of $CaCO_3$	No potential for scaling. $CaCO_3$ will dissolve.
0	pH of water = pH_s of $CaCO_3$	Stable water. If the temperature of water or other factors change, scaling may occur.

As the pH is decreased the CO_3^{2-} component of alkalinity decreases reducing the driving force for $CaCO_3$.

LSI	$= 2.5$
pH of cooling water	$= 9$
Therefore pHs	$= 9.0 - 2.5 = 6.5$
pH > pHs	$CaCO_3$ scaling is likely.

Acid addition will depress the pH.

5.6.1.2   What the LSI is Not

The LSI does not predict the corrosivity of water. Corrosion may still occur due to dissolved oxygen and dissolved solids such as chlorides. It does not predict the scaling potential of other salts.

The indices provide only the thermodynamic *driving force* for scaling – not the quantity of $CaCO_3$ – that will actually precipitate. In the presence of crystal-modifying chemical inhibitors, these indices are not accurate [6].

5.6.1.3   Scale-control Methods

A number of methods exist to control scale formation. These are

- acid addition to reduce the pH
- increasing the solubility of scale-forming substances
- removing calcium hardness from the water
- precipitating the scale-forming substances.

(i) Acid addition

Sulphuric or hydrochloric acid addition was the traditional method to reduce scale control by depressing the pH as per Equation (5.16).

$$H_2SO_4 + Ca(HCO_3)_2 \longrightarrow CaSO_4 + 2CO_2 + 2H_2O \qquad (5.16)$$

Given that calcium sulphate ($CaSO_4$) is 50 times as soluble as $CaCO_3$, sulphuric acid addition is commonly used as an inexpensive method in power plants to control LSI. However, this introduces another potential scalant and a safety hazard.

(ii) Chemical inhibitors

In industrial cooling systems scale inhibitors such as crystal modifiers and dispersants are able to handle high hardness waters without the risks associated with acid handling [6–8]. These compounds include phosphonates and polymers such as polyacrylate or its co-polymers. Crystal modifiers act by distorting the $CaCO_3$ crystal structure – and dispersants act by reinforcing the negative charge present on the surface of colloidal matter – to prevent them binding together.

Calcium can be removed by lime soda softening, ion exchange or membrane filtration (nanofiltration or reverse osmosis).

5.6.2   *Fouling*

Fouling occurs due to suspended solids in the cooling water forming deposits on the system's surface.

Suspended solids can be

- air-borne substances from exhaust vents located nearby
- sand
- biofilms
- aluminium floc
- iron oxides and other corrosion products
- process leaks such as hydrocarbons.

Fouling prevents biocides and corrosion inhibitors to reach the micro-organisms and the metal surface. Control of fouling is now more important than ever, given that cooling systems are operated for longer periods between cleanings.

Fouling-control strategies are

- sidestream filtration of cooling water
- increasing cooling water velocities
- addition of dispersant chemicals (1–10 mg/L as active ingredient) to reinforce the negative charge of suspended materials.
- injecting rubber balls to maintain clean heat exchanger tubes
- equipment redesign to remove low flow areas.

5.6.3   *Corrosion*

Corrosion can be defined as the destruction of a metal by chemical or electrochemical reaction with its environment resulting in the formation of metal compounds that decrease the structural strength of the metal. Therefore corrosion contributes to equipment failure, plant downtime and decreased heat transfer due to accumulation of corrosion products.

The most common types of corrosion in cooling systems are

- General corrosion
- Pitting corrosion
- Galvanic corrosion
- Stress corrosion cracking
- Dezincification.

General corrosion is a uniform attack over the entire metal surface. Corrosion products are voluminous and contribute to fouling and under deposit corrosion.

Pitting corrosion is far more serious than general corrosion since the attack occurs over a small localised area where overall metal loss is negligible and can lead to heat exchanger failure. Chlorides concentrate in the pit and produce an acidic environment with low oxygen concentration.

Galvanic attack occurs when two different metals are in contact with one another. Common examples are steel pipes screwed into a bronze valve. The steel pipe is the anode and the bronze valve is the cathode. The anode corrodes faster than the bronze valve.

Stress corrosion cracking in cooling systems is due to chlorides and is most prevalent in austenitic stainless steels (304 and 316). **Dezincification** is where the zinc is preferentially removed from a copper alloy leaving the metal brittle.

Figure 5.12 shows a series of carbon steel corrosion coupons with varying corrosion rates.

The principal factors governing corrosion are shown in Table 5.7.

Corrosion control in cooling water systems is achieved by

- selecting non-corrosive metals
- application of protective coatings
- application of sacrificial anodes
- adjust pH
- dosing of corrosion inhibitors such as anodic, cathodic and filmforming corrosion inhibitors. Examples are ortho-phosphate (anodic), zinc (cathodic) and tolyltriazole (film forming).

Table 5.8 gives guidelines for assessing corrosion rates.

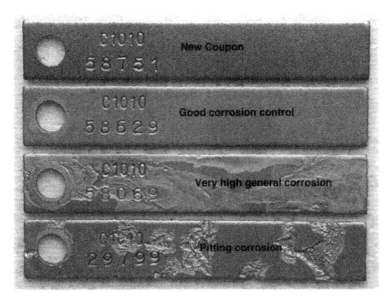

Figure 5.12   Corrosion coupons illustrating the types of corrosion

Courtesy of Absolute Water Treatment Pty Ltd.

Table 5.7　**Principal factors contributing to corrosion**

Factor	Comments
pH	Acidic and slightly alkaline water dissolves metal and the protective oxide film on metal surfaces. Alkaline water favours the formation of the protective oxide layer.
Dissolved oxygen	Oxygen is essential for the cathodic reaction to take place.
Temperature	Below 71° C (160° F) for every 13–27° C (25– 50° F) increase in temperature causes the corrosion rate to double.
Dissolved salts	Cl and SO_4 increases the conductivity of water and increases corrosion rates. A measure of corrosion potential is the Larson – Skold indicator.
Velocity	High-velocity water increases corrosion by transporting oxygen to the metal surface and by carrying away corrosion products at a faster rate. It also causes erosion of metal surfaces. Specified water velocity for carbon steel is 1.0–1.8 m/s and for 90/10 copper nickel 1.0–2.5 m/s.
Microbial growth	Microbial growth promotes microbiologically influenced corrosion (MIC) due to the formation of corrosion cells. By-products of anaerobic bacteria forms acids.
Process leaks	Process leaks can lead to increased sulphate-reducing bacterial growth which causes corrosion.

Table 5.8　**Guidelines for assessing corrosion**

Metal	Corrosion Rate		Comment
	µm/y	mpy*	
Carbon steel	0–50	0–2	Excellent corrosion rate
	50–75	2–3	Generally acceptable for all systems
	75–125	3–5	Moderate corrosion ()
	>125	>5	Unacceptable. Iron fouling is likely
Copper alloys	0–5	0–0.2	Acceptable corrosion rates
	5–12.5	0.2–0.5	Corrosion rates too high. Can cause copper plating of mild steel exchangers.
	>12.5	0.5	Unacceptable
Stainless steel	0–25	0–1	Acceptable
	>25	1	Unacceptable corrosion rates

mpy* –mils per year (one thousandth of an in.)

5.6.4　*Biofouling and Microbial Growth*

Cooling towers and evaporative condensers are highly effective scrubbers. Temperature and pH are ideal for the growth of micro-organisms in the presence of plenty of nutrients. Contributing to this favourable environment is sunlight and oxygen. In once-through sea water or brackish water systems biofouling is due to macrofoulants such as zebra mussels or Asiatic clams or similar species. Macrofoulants are not considered in this section.

Biofilms due to micro-organisms are undesirable because they plug nozzles, decrease flow rates, hinder heat transfer and contribute to under deposit corrosion. A 1 μm (0.00004 in.) thickness biofilm on a stainless steel tube is equivalent to a 25 μm increase in tube thickness [11]. The thermal conductivity of biofilm is approximately 0.65 W/m-K. In contrast, metals have conductivities between 16 and 384 W/m-K. Therefore they are 25–600 times more resistant to heat transfer than metals. In a power plant this difference translates to thousands of dollars lost per hour.

There are three types of micro-organisms of interest:

1. Algae – found in distribution decks of cooling towers
2. Fungi – contributes to wood decay
3. Bacteria – causes fouling, Legionnaires' disease and corrosion.

5.6.4.1   Microbiological Control

To combat microfouling biocides are added to the cooling water. Biocides act to decrease the rate of microbiological activity by killing the organisms through variety of ways. Biocides addition need to form part of a holistic microbiological control strategy.

Legionella control is one such example. Many organisations and regulatory bodies have developed guidelines [13–17]. Specific action on *Legionella* control could be to develop a risk management plan in line with the requirements of your local regulatory body for the control of *Legionella*. The Victorian Department of Human Services, Australia [13] and Standards Australia [14] have developed such risk management plans.

The following are the three types of disinfection methods:

- Oxidising biocides such as chlorine, sodium hypochlorite, bromine, ozone, organic halogen donors, chlorine dioxide, peroxides and monochloroamine. They are non-specific and have broad-spectrum biocidal activity. The active ingredient in halogen chemistry is hypochlorous (HOCl) or hypobromous (HOBr) acids. HOBr is more effective at pH above 8. They dissipate in the cooling tower and for this reason they are added continuously or slug dosed daily. Halogen-based products have a residual effect and are added to maintain a free chlorine residual of 0.5–1.0 mg/L. Halogens form trihalomethanes (THM), a controlled substance, and for this reason halogenation is under scrutiny.
- Ultraviolet disinfection at a wavelength of 253.7 nm at a minimum dosage of 30 000–40 000 μW – s/cm^2 is also effective [15]. However, it also suffers from not having a residual effect.
- Non-oxidising biocides (made from complex organic molecules) are much more specific to one or more organisms, slow to react and some compounds degrade to benign substances. They need to be dosed

in large concentrations, are persistent and relatively expensive. The most common are isothiazolones, 2,2-dibromo-3-nitrilopropionamide (DBNPA), glutaraldehyde and quaternary ammonium compounds. The last compound also has surfactant properties and is therefore effective against sessile bacteria.

Typical non-oxidising biocide dosages are

Isothiazolinone (1.5%)   75 mg/L
Glutaraldehyde (45%)   100 mg/L
DBNPA (20%)   40 mg/L.

When dosing biocides, it is important to consider the holding time index (HTI) of the biocide in the cooling water. The HTI is the amount of time that it takes a biocide to halve its concentration. High bleed rates will make the biocide ineffective once it decreases below its effective dosage. The following calculation best illustrates this point.

Holding Time Index is given as $HTI = 0.693 \times volume/blowdown$ rate

Lets assume that the biocide was dosed at 100 mg/L and is expected that the biocide addition is done once a week. The minimum concentration required is 75 mg/L. The volume of the cooling system is 1000 m^3. Blowdown rate at 2 cycles of concentration is 50 m^3/hr and at 4 cycles of concentration is 17 m^3/hr. The effective HTI values for these two scenarios are

HTI at 2 cycles of concentration = 14 hrs
HTI at 4 cycles of concentration = 42 hrs.

Thus at high blowdown rates (2 cycles of concentration) there is wastage of biocides since it has been depleted in less than a day to below its effective dosage.

Other possible actions to control microbiological organisms are

- eliminate plumbing 'dead-ends' or modify closed loops
- select the biocides based on system half-life and water chemistry (isothiazolones, glutaraldehydes are effective for *Legionella* control)
- use a mixture of oxidising and non-oxidising biocides
- prevent build up of solids in the cooling tower sump
- shield wetted areas from sunlight if practical.

Guidelines for microbial monitoring are given in Table 5.9. *Legionella* testing needs to be carried out by specialised laboratories. Total plate counts need to be based on agar petri dish method rather than dip slides which are unreliable.

Table 5.9 Guidelines for microbial control

Total plate count cfu/ml	*Legionella* count CFU/mL	Comments
0–10,000	<10	Effective microbial control
10,000–10^5	10–100	General microbial control is satisfactory but *Legionella* control is unsatisfactory. Increased monitoring required. Increase biocide dosage if required.
>10^5	100–1000	Potentially hazardous situation. Re-evaluate microbial control programme.
>10^5	>1000	Serious situation. Shutdown immediately and decontaminate.

Cfu/mL = colony-forming units per millilitre.

5.7 Role of Water Treatment Contractors in Water Conservation

Most cooling water-treatment programmes are outsourced to water treatment companies. Therefore gaining their commitment is critical to the success of the water conservation programme. A knowledgeable cooling water specialist can be of immense assistance to making the water conservation programme a success. Therefore, contracts for cooling water treatment contracts should never be awarded on the *lowest price* as the sole criterion.

Below are some suggestions to get the best from your cooling water contract and vendor.

- The cooling towers need to be operated at the highest practical cycles of concentration. For this to happen, the contract needs to stipulate the desired number of cycles. Use this as a key performance indicator (KPI).
- Cooling water contracts need to stipulate a 'management fee' and a 'performance fee'. The management fee is to meet the basic contractual terms such as on-time delivery, dosing pumps maintained and reporting of trends. The performance fee is to ensure that the KPIs for water consumption, system chemistry, corrosion, microbial levels are met or exceeded. The minimum analytical requirements need to be stipulated in the contract.
- The make-up water flow and blowdown water flow and conductivity should be monitored, logged and charted. The cycles of concentration should be logged and charted and reasons given for any variations.
- Acceptable corrosion rates should be included in the contract. Corrosion monitoring should be reported at 30-, 60- and 90-day intervals. The corrosion coupon rack should be sited at the outlet of the hottest heat exchanger, if accessible. Otherwise the results may not be

representative. It is a good idea to stipulate the corrosion inhibitors for the metallurgy present in the system.

- Some vendors – when asked to raise cycles – bring in the spectre of *Legionella*. Increasing cycles of concentration has nothing to do with *Legionella* control. Be wary of such vendors. It is more a case that their water-treatment chemicals are incapable of handling stressed conditions.
- During site visits to the facility, the vendor should examine the water systems for indications of leakage, overflows and other types of water loss (or other deleterious conditions) and report them promptly for corrective action. Any overflows in the cooling tower, problems or leaks need to be promptly reported to the customer.
- The chemical treatment programme must specify the ingredients not the trade names (most are meaningless), dosage and active % used for corrosion, scaling and microbial treatment, so that chemical treatment programmes can be compared against each other or should be based on $ to treat 1 m³ of blowdown water.
- Regular quarterly reviews with plant manager are needed to be held to assess performance. Statistical control packages are available these days to chart all the variables and this will show the percentage of time that the key variables were within the specified limits. They tell a better story.
- Use of alternative water sources should be assessed by the supplier. The performance fee needs to be tied to this. Computer programs such as Water Cycle™ are available for this purpose.
- Provide instructions and 'hands on' training to plant and building engineers as required so that water-treatment programme instructions, tasks and remedial actions for out-of-control situations are understood.

5.8 Conclusion

We have examined the types of cooling towers, the water conservation opportunities both voluntary and non-voluntary, alternative water sources, factors to consider in water treatment of cooling water systems and finally the role water-treatment specialists in managing water use in cooling water systems.

References

[1] Australian Institute of Refrigeration, Air conditioning and Heating. *Last Days of the Cooling Tower*. EcoLibrium. pp. 8–9. March 2005.
[2] Sydney Water. *Water Conservation Best Practice Guidelines for Cooling Towers in Commercial Buildings*. 2004.

[3]   Burger R. *Colder Cooling Water Pays Off*. Chemical Engineering. McGraw Hill Inc. March 1991.

[4]   North Carolina Department of Environment and Natural Resources, Division of Pollution Prevention and Environmental Assistance. *Case Study: Duke University*. July 2003.

[5]   O'Neill C.R. *Impacts and Control of Zebra Mussel*, IWC 99–04. Proceedings of the International Water Conference. Pittsburgh. 1999.

[6]   Puckorius P.R. and Loretitsch G.R. *Cooling Water Scale and Scale Indices: What They Mean – How to Use Them Effectively – How They Can Cut Treatment Costs*, IWC 99–47. Proceedings of the International Water Conference. Pittsburgh. 1999.

[7]   Strauss S.D. and Puckorius P. *Cooling Water Treatment for Control of Sscaling, Fouling, Corrosion*. Power. June 1984.

[8]   Buecker B. *Power Plant Water Chemistry – A Practical Guide*. Pennwell Publishing Inc.: Tulsa. 1997.

[9]   Kirby G.N. *Corrosion Performance of Carbon Steel*. Chemical Engineering. March 1979.

[10]   Tuthill A.H. *The Right Metal For Heat Exchanger Tubes*. Chemical Engineering. 1990.

[11]   Ascolese C.R. Technical Paper 322, "Biofilm Formation: Attendant Problems, Monitoring, and Treatment in Industrial Cooling Systems". 27th Annual Liberty Bell Corrosion Course. Philadelphia, PA. 1990.

[12]   Victoria Department of Human Services. *A Guide to Developing Risk Management Plans for Cooling Tower Systems*, www.legionella.vic.gov.au. 2001.

[13]   Standards Australia AS 5059. *Power station cooling tower water systems – Management of legionnaire's disease health risk*. 2003.

[14]   NSW Health Department. *NSW Code of Practice for the control of Legionnaires' Disease*. 2nd edition, February 2002.

[15]   Gump D.J. *Utilisation of Ultraviolet disinfection systems in industrial processes*. Severn Trent Services Inc. Ultra Dynamics. Colmar, PA. 2001.

[16]   Cooling Technology Institute *Legionellosis – Guideline: Best Practices for Control of Legionella*. February 2000.

[17]   American Society of Heating, Refrigerating And Air-Conditioning Engineers Inc. ASHRAE Guideline 12–2000 *Minimising the Risk of Legionellosis Associated with Building Water Systems*. 2000.

[18]   Meyer W.C. *Avoid Legionellosis Lawsuits over Cooling Towers*. Chemical Engineering. McGraw-Hill Book Company. September 2000.

[19]   Puckorius P.R. *Cooling Tower Systems & Legionella Bacteria: Recent USA Incidents/Outbreaks/Causes/Corrective Action Methods – Are We Making Progress? What Action Should be Taken*, IWC 99–17. International Water Conference. Engineers Society of Western Pennsylvania, October 18–20, 1999.

Chapter 6

Alternatives to Wet Cooling Towers

The previous chapter was devoted to understanding and taking control of the cooling system. This chapter will discuss strategies to minimise the cooling load in air-conditioning and refrigeration systems, which is the main requirement for cooling water in these systems. By improving energy efficiency of air-conditioning and refrigeration systems both energy and water savings can be realised. It will discuss alternatives to wet cooling systems; in particular, air-cooled chillers/condensers, hybrid cooling systems, combination of air-cooled and wet cooling systems and geothermal applications. Their suitability will depend on individual circumstances such as adequate land area.

6.1 Air-conditioning and Refrigeration Systems

Before a discussion on energy saving options can begin, a basic understanding of air-conditioning system is required. A refrigeration system has many of the components of an air-conditioning system. Air-conditioning and refrigeration can be achieved by mechanical and absorption refrigeration systems. Absorption refrigeration systems are economical when there is a source of cheap energy such as waste heat. However, most facilities do not have access to this and therefore absorption systems are not discussed in this chapter. Chillers are classified as reciprocating, screw and centrifugal depending on the type of compressor used. Reciprocating chiller compressors are used below 700 kWR (200 tons) and screw compressors are used in the range from 140 to 2800 kWR (40 to 800 tons). Centrifugal compressors are available from 263 kWR (75 tons), but are typically used in large installations and can exceed 17 585 kWR (5000 tons) or more. The centrifugal compressor is ideal for air-conditioning systems since it can operate at variable loads, has few moving parts and is economical to run. Figure 6.1 shows a schematic of a typical air-conditioning system and Figure 6.2 shows a photograph of centrifugal compressor and chiller.

The air-handling units circulate a mixture of fresh and recycled air. They transfer the heat contained in a building from air to the chilled water and

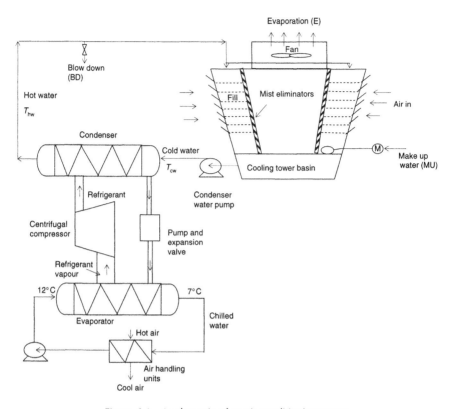

Figure 6.1   A schematic of an air-conditioning system

in the process increasing the chilled water temperatures from 7°C to 12°C (45°F to 54°F). The chilled water enters the evaporator (a shell and tube heat exchanger) where the 'hot' chilled water transfers the heat to the liquid refrigerant. The refrigerant evaporates and in the process creates the cooling effect for the chilled water. The vapour enters the compressor where it is compressed and in the process picks up heat. The high-pressure superheated vapour flows into the condenser (a shell and tube exchanger) where it gives up the heat to the circulating cooling water and in the process is condensed into a saturated liquid. The condenser water temperatures rise from 28° C–32° C (80° F–90° F) to 32° C–38° C (90° F–100° F). The condenser water dissipates the heat in the cooling tower. The liquid refrigerant is further sub-cooled in the expansion device and flows backs into the chiller for the cycle to repeat again.

6.2   Energy Conservation = Water Conservation

Process cooling and air-conditioning can account for a significant portion of the organisation's electricity bill. For instance, in a brewery approximately 32% of electricity is consumed in the refrigeration system. It is the second largest energy user after pumps and compressors. Therefore by reducing

Figure 6.2 A photo of a centrifugal compressor and chiller

energy consumption, the bottom line improves, reduces water usage and reduces greenhouse gas emissions.

Ways to minimise energy consumption in air-conditioning and refrigeration systems are summarised in the following rules.

Rule 1: Operate the system only when needed

In comfort cooling applications, air-conditioning systems including compressors and cooling towers should be operated only when areas are occupied. Many buildings are unoccupied during weekends and after 5PM, yet the air-conditioning systems are kept operating in buildings that lack a comprehensive building management system. In process applications use the refrigeration system only on a need-to-use basis. Eliminate non-essential heat loads.

Improved temperature control of process temperatures, cooling water temperatures and cooling water flow rate lead to optimal operation and efficiency.

In batch-operating plants it is important to shut off cooling water to process equipment once the cooling needs are met. A common occurrence in plants with manual control of cooling water flows is that the cooling water pumps operate long after the process has finished. With automatic temperature control this is easily avoided.

Rule 2: Eliminate over-cooling

Over-cooling can be eliminated by revising operating procedures and modifying air-conditioning system controls. Rather than maintaining a constant temperature, allow the temperature to fluctuate within a dead ban range.

Rule 3: Carryout timely maintenance

It is often possible to gain energy efficiency improvements at a very low cost by carrying out timely maintenance of the refrigeration and ancillary equipment. Simple things such as shutting doors and cleaning of condenser and evaporator units and timely removal of scale improve heat transfer. It is estimated that 3 mm of scale will increase power input by 30% and reduce power output by 20%.

Rule 4: Reduce lighting load

Lighting in a typical office building accounts for 40% of the energy demand. Whilst in industrial plants lighting accounts for about of 5–7% of the energy demand, it is still a sensible measure to reduce energy usage where practical.

Less lighting = less heat generated = less air-conditioning load

In office buildings it is claimed that savings of 30% are achievable without impacting on use or comfort. The improvement measures can take the form of switching off lighting when not required or installing movement sensors to switch on lighting, replacing energy-hungry lighting with more efficient types such as fluorescent Triphosphor T5 (23% more efficient than a triphosphor T8 lamp [1]) and designing new buildings to have more natural light by having double-height windows and curved ceilings. As a guide, lighting levels in office buildings above $11 \, W/m^2$ is considered to be energy inefficient [1].

Case Study: Louis Stokes Laboratories, National Institute of Health

The Louis Stokes Laboratories of the US National Institutes of Health applied a combination of strategies to improve the efficiency of lighting systems. First, all fluorescent light fixtures were replaced with more efficient T-8 lights and electronic ballasts. Secondly, motion sensor-activated lights and light emitting diode (LED) exit sign were installed. Thirdly, a programmable lighting control system was installed. Lastly, double-height windows and curved ceilings were employed to allow more daylight into the lab space. Based on these measures, the average energy consumption rate per unit of lighting area was reduced to an impressive $17 \, W/m^2$ ($1.6 \, W/gross \, ft^2$) [2].

Rule 5: Increase building reflection

Designing the building to reduce heat from solar radiation reduces the air-conditioning load especially in sunny hot climates. Experiments in California and Florida have demonstrated that coating roofs with a reflective coating reduces summertime average daily air-conditioning electricity use from 2% to 63% [3]. Two medical offices in Northern California reduced air-conditioning loads by 8 and 12% respectively. For colder climates the energy savings may not be that dramatic since in winter these savings can be negated by the added heating requirements.

Roof gardens provide both cooling and heating benefits. In Germany roof gardens on top of office buildings is quite popular.

Rule 6: Improve building insulation

Building insulation reduces both cooling and heating loads in buildings. Refer to the State or National standards.

Rule 7: Install energy efficient chillers and refrigeration systems

Install energy efficient chillers and refrigeration systems. Modern centrifugal chillers use half the energy usage of models that were installed a decade ago. Potential saving: 1.2% of a facility's total energy use with an average payback of 23 months [4].

Rule 8: Size chillers to better balance the load

Sometimes one large and another small chiller may be more efficient than one large chiller because at part load operation it is less efficient. The cost savings can be significant.

Rule 9: Use the free cooling mode – economiser cycle

Many air-conditioning systems operate with a fixed amount of outdoor air. The mechanical refrigeration load can be reduced by modifying the system to use natural ventilation using cooler ambient air at up to 100% of its supply airflow, when the ambient air is cooler than the return air. This is known as the "economy air cycle". Dampers are opened according to the outdoor dry bulb temperature or by sensing the enthalpy difference between the outdoor and the indoor enthalpy of air. Not operating the air-conditioning system results in energy and water savings. The overall economics, frequency of operation and control logic required needs to be evaluated before going ahead with this option. Energy savings as much as 40% depending on location and load profile can be realised [4].

On the other hand reducing cool room exhaust air flows during summer reduces energy consumption.

Rule 10: Operate at the lowest condensing water temperature

A common strategy to reduce energy usage is to reduce the condensing water temperature. Cooling towers are designed for summer month operation, and therefore during part load operation and in winter months it is possible to reduce the condensing water temperature thus reducing the condenser refrigerant pressure. For example, Table 6.1 shows that a reduction in condenser water temperature from 29.4° C to 21.1° C (85° F–70° F) reduces chiller energy consumption from 0.162 to 0.128 kW/kWR (0.57–0.45 kW/ton), a reduction of 21% [4].

Rule 11: Operate at the highest chilled water temperature

By increasing chilled water temperature by one degree, chiller energy usage can be reduced by 0.6–2.5% [4]. Whilst it increases the power input to the compressor, the gain in refrigeration output is much greater. The efficiency gains are more dramatic at the lower set point temperatures than at higher temperatures. In process applications this may not always be possible. However, the opportunity to do this arises when chilled water flows are throttled indicating that the temperatures are lower than required.

Rule 12: Install variable drive speed fans to the cooling tower

Cooling towers are sized so that they have sufficient heat rejection capacity during summer when the hottest ambient temperatures are expected. When the actual wet bulb temperature is below the design wet bulb temperature or where the heat load appears to be lower, the designed amount of cooling can be obtained with lower airflow rates. As the airflow rate decreases, the fan speed and the motor power requirements also decrease. A variable speed drive fan controlled by the cooling water inlet temperature can decrease both water and energy costs. It is estimated that a saving 30–60% decrease in cooling energy use can be realised depending on load profile [5, 6]. Processes with variable cooling loads such as in batch production systems and those with spare machine capacity are likely to make major savings.

Table 6.1 Effect of condenser water temperatures on chiller energy usage

Cooling water temperature		Typical chiller energy consumption		Energy savings
°C	°F	kW/kWR	kW/ton	
29.4	85	0.162	0.57	Base (%)
28.3	83	0.154	0.542	5
26.7	80	0.149	0.524	8
23.9	75	0.138	0.484	15
21.1	70	0.128	0.45	21

> **Case Study: Chemical Manufacturer UK**
>
> A manufacturer of synthetic drugs operates a batch production system which leads to varying cooling loads. After examining its energy usage the company installed a frequency inverter variable speed drive to match fan speed to actual cooling load which resulted in a 60% reduction in electricity consumption.
>
> Adapted from: *Energy Efficiency Best Practice Programme. Case Study 270. Variable Speed drive on a cooling tower induced draught fan.*

Rule 13: Recover waste heat

Another effective way to reduce heat rejection, and hence water consumption, is by recovering and re-using the heat that would otherwise be rejected to the environment. Many water-cooled chillers can be purchased with a heat recovery unit, and depending on the type of refrigerant, the condensing pressure, the design and the application, a significant portion of the waste heat can be recovered. For every kW of heat recovered in a water-cooled plant, up to 1.7 L of water can be saved per hour of operation. Such recovered waste heat can be used in many different ways, such as

- Low grade heat at 30° C–40° C (86° F–104° F) for space heating, reheat of air in air-handling units and as boiled feed water.
- High grade heat – hot water 50° C–70° C (122° F–158° F) as hot water for wash down.

6.3 Alternative Heat Rejection Systems

There are number of alternative heat rejection systems, other than evaporative condensers and cooling towers. Fluid coolers, chillers and condensers can be supplied in air-cooled or in various hybrid cooling designs to reject heat to the environment with minimal water usage. Another option is to use a combination of evaporative and air-cooled equipment. Where there is an excess of low-grade waste heat, absorption chillers can be used either as a stand-alone unit or in combination with electric chillers. The strategy is to use the absorption chillers preferentially during high electric rates.

Then there are alternatives that avoid direct rejection to the environment, such as geothermal systems. The best time to consider these alternatives is when designing a new system or upgrading/replacing an existing cooling system. These technologies are described briefly.

Figure 6.3   A bank of air-cooled condensers

Courtesy of Minus 40 Pty Ltd.

6.3.1   Air-Cooled Condensers

In air-cooled condensers, the refrigerant condenses inside finned tube over which a stream of air flows. The cooling tubes/coils have fins to increase heat transfer. The air is blown by fans and cooling takes place due to conduction and convection. Figure 6.3 shows a bank of air-cooled condensers.

The advantages of air-cooled chillers and condensers are well known. They do not require cooling water and therefore the associated maintenance issues of corrosion/scaling, risk of *Legionella* and chemical treatment costs are eliminated.

However, air-cooled condensers pay a penalty in terms of

- Unlike cooling towers which use the wet bulb temperature as a heat sink, the end process temperature achievable with air-cooled condensers is higher since air cooling uses dry bulb temperatures as the heat sink. Table 6.2 shows that typically the dry bulb temperatures are

Table 6.2 **Wet and dry bulb temperatures for some selected European cities**

City	Wet bulb temperature		Dry bulb temperature	
	° C	° F	° C	° F
Athens	22	71.6	36	97
Berlin	20	68	29	84
London	20	68	28	82
Paris	21	69.8	32	90
Rome	23	73.4	34	93

8°–14° more than the corresponding wet bulb temperature. They also have a higher approach temperature range of 10° C–15° C

In practice, air-cooled condensers are used in manufacturing plants when the process temperatures are greater than 70° C–80° C, such as in oil refineries. Given these constraints, air-cooled systems are less efficient than comparable wet cooling systems.

- As mentioned in Chapter 5, in cooling towers the primary cooling mechanism is due to the evaporation of water given the high latent heat of water (2431 kJ/kg). Sensible cooling only plays a secondary role. However, in air-cooled systems, only sensible cooling can be used for cooling. Moreover, the specific heat capacity of air is four times as low (1 kJ/kg K as against water 4.2 kJ/kg K). It also suffers from a lower heat transfer coefficiency. Therefore, for the same heat transfer duty, air-cooled systems require more air flow and larger heat transfer area. For this reason they require a greater fan capacity to deliver more air and more space.

Despite these disadvantages, air-cooled systems still have a place in cooling systems. The decision to install an air-cooled chiller or condenser needs to be considered by giving due consideration to the following:

- minimising energy usage
- minimising heat emissions
- minimising plume emissions
- minimising noise emissions
- minimising water discharged to the environment
- minimising the risk of *Legionella*.

In some cases other factors may override the relatively higher cost and energy inefficiency of air-cooled chiller/condensers such as the risk of *Legionella*.

Sydney Water commissioned a study [6] to ascertain the breakeven point for the operation of air-cooled chillers versus water-cooled chiller for air-conditioning systems in commercial buildings. Figure 6.4 shows that over a 25-year life cycle, air-cooled systems were technically and financially economical at cooling loads below 450 kWR (128 Ref Tons). Archibald [7] who looked at a number of scenarios arrived at similar conclusions.

6.3.2   Hybrid Cooling Towers

Hybrid cooling towers (HCTs) utilise the best features of both air-cooled and water-cooled systems, selecting the most water- and energy-efficient mode to suit the operating and ambient conditions.

Another advantage of HCTs over cooling towers is that they eliminate the issue of plume when operating in the dry mode. This is especially advantageous in populated areas where plume formation is popularly (and falsely) associated with energy wastage or air pollution.

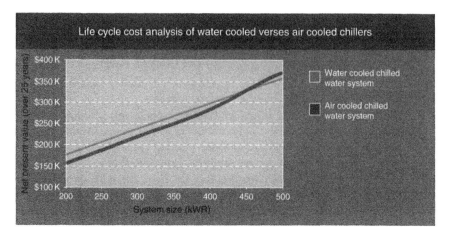

Figure 6.4   Life-cycle analysis of water-cooled and air-cooled chillers

Courtesy of Sydney Water, *Best Practice Guidelines for cooling towers in commercial buildings*, 2004.

Several competing hybrid concepts are now available on the market, each with its own advantages and disadvantages. Most can operate in two or three modes. The different modes of operation are as follows:

- **Dry mode**, in which the HCT operates essentially as an air-cooled condenser or cooler, with no water usage. Heat is rejected by warming of the air, that is only sensible cooling takes place.
- **Evaporative pre-cooled mode**, in which water is used to pre-cool the air before it reaches the heat exchange surface either as spray cooling or as closed adiabatic cooling. However, except during low humidity ambient conditions this mode can consume more water than conventional cooling water systems, at comparable cooling water or condensing temperature conditions.
- **Evaporative cooling mode**, in which the water is used to wet the surface of the heat exchange surface directly. Heat is rejected by evaporation of the water in contact with the heat exchanger. In this mode it acts as a conventional cooling tower.

In the dry/wet mode both sensible and evaporative (latent heat) heat transfer modes are used. During the cooler periods only the dry cooling tower section is used. Compared to conventional evaporative cooler significant water savings can be obtained at peak conditions.

These systems are generally more expensive than conventional air-cooled condensers/fluid coolers. Therefore, a life-cycle costing needs to be carried out to assess their cost savings over a 15-year period.

Figure 6.5 shows the Baltimore Aircoil HXI cooler which also incorporates a finned coil section at the top of the tower to eliminate plumes.

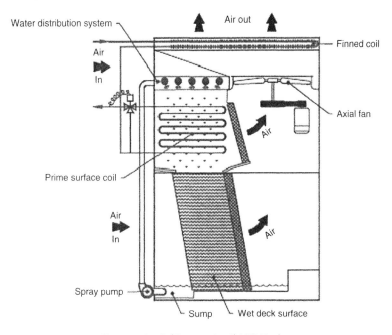

Figure 6.5   Baltimore Aircoil HXI Cooler

Courtesy of Baltimore Aircoil.

Figure 6.6 shows a cost comparison between a HXI cooler, which uses the combined technology, and that of conventional open recirculating cooling tower. While the capital cost is about four times that of a conventional cooling tower, the operating cost is only a fraction of conventional technologies.

An important consideration in the selection of a hybrid cooling system is the switch point, which is defined as the maximum air temperature at which design cooling capacity can be achieved with dry operation. For Sydney conditions a low switch point of some models generally below 14° C will dictate wet operation for a substantial portion of the year. Therefore a higher switch point 22° C gives a much longer dry operation window, and potentially greater water savings.

The HCTs systems do have operational issues such as the need to control micro-organisms through use of biocides; some models may scale up with time. To overcome the scaling problem, some designs incorporate a reverse osmosis membrane plant to spray only demineralised water on to the coils.

6.3.3 Combination Cooling Systems

Under certain conditions, a combination of evaporative and air-cooling can present a technically and commercially viable alternative to the use of either technology alone.

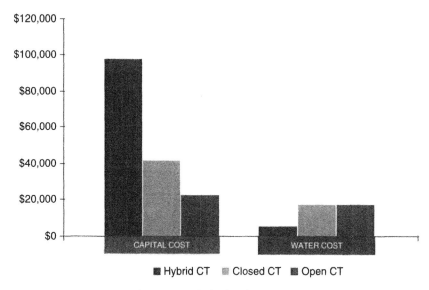

Figure 6.6   Cost comparison between HXI hybrid cooling towers and conventional wet cooling systems [8]

Comparison is based on a cooling load of 2030 kW, an inlet temperature of 40° C, an outlet temperature of 30° C and a wet bulb temperature of 24° C.

Courtesy of Baltimore Aircoil.

Generally evaporative cooling, be it an evaporative condenser or a cooling tower, is a low-capital and energy-efficient form of heat rejection. Air-cooling and especially dedicated hybrid cooling systems are far more costly. In many cases installing separate air-cooling systems in parallel to conventional evaporative cooling systems offers both energy and water savings by operating the air-cooled systems during low-ambient temperature and low-load conditions, and the evaporative during high load and high ambient only.

An analysis of typical Sydney ambient conditions shows that approximately 60% of the year, the ambient dry bulb temperatures are below 18° C, and hence well suited to air-cooling. Considering that in many cases a large proportion of the remaining 40% of the year would coincide with low load conditions, this would make air-cooling viable for as much as 80% of the year, whilst the remaining 20% would be served with evaporative cooling, with significant water savings and little or no energy penalty.

Several case studies have shown that in many cases the installation of separate air-cooled and evaporative condensers can be cheaper than a single hybrid unit only, whilst offering similar water and energy savings.

Another attractive advantage of combination systems is the redundancy achieved through the installation of dual systems. In the event that either system should fail, the other can be run, albeit at a water or power consumption penalty, until repairs have been undertaken.

The use of a combination of air-cooled and water-cooled systems has the following disadvantages:

- Significant space is required to install both systems. This is not often available.
- An intelligent control system, programmed to suit the specific design of the plant, is required to ensure that the optimum mode of operation is selected at all times.

Case Study: Red Lea Chicken, Sydney, Australia

Red Lea Chicken is a family-owned business that started in 1957. It processes in excess of 300 000 chicken a week. They decided to upgrade their system and opted for air chilling rather than use a spin chiller. This required additional cooling load. Instead of installing an evaporative condenser, Red Lea decided to install a large air-cooled condenser thus obviating the need for water use. They also installed water sprays on the condenser to cater for those hot days > 45°C (113°F) without the need for pre-cooling pads; installed variable speed control for the fans as well as a heat exchanger to recover waste heat up to 600 kW at a temperature of 60°C (140°F) that can be used as wash down water.

The decision to use air chillers resulted in a water saving of 200 m³/day.

Adapted from: Sydney Water. *Red Lea returns on $3 million investment. The Conserver.* Issue 10. May 2006.

6.3.4 Geothermal Cooling Systems

In geothermal cooling systems, the earth is used as the heat sink. Whilst ambient air temperatures vary widely over a calendar year, 2 m below the earth's surface, the temperatures are fairly constant at 10°C–15°C range (50°F–60°F). Consequently, the earth is cooler than ambient temperatures during the warmer months and in winter it is warmer than the ambient temperature. This aspect is used in geothermal systems. In summer, the ground loop acts as a 'heat sink', rejecting the unwanted building heat. In winter, it acts as a 'heat source', absorbing heat from the ground to heat the building.

Geothermal cooling systems consist of piping loops, pumping system, condenser and air-handling units. The piping loops are filled with water and an antifreeze solution. Cooling loops made from high-density polyethylene are installed either vertically or horizontally depending on the geological formation and available land area.

If adequate land area is available without hard rock formations, then horizontal piping loops are more cost effective. These are found in open fields or under car parks. In the horizontal loop system, trenches are dug to an average depth of 1.2–1.8 m (4–6 ft). The length of the loop piping is dependent on ground temperature, thermal conductivity of the soil, soil moisture and system design.

In commercial buildings or educational facilities, vertical loops are more common due to lack of space, or is too rocky. To install a vertical loop, holes are bored 45–76 m (150–250 ft) into the ground. Long, hairpin-shaped U loops of pipe are then inserted. The hole is backfilled and cemented, the pipes are connected to headers in a trench leading back to the building. The drilling depth is determined by the lowest total cost based on the conditions at the job site. If the depth required is excessive, then multiple loops are used. A bore might have a heat rejection of 6 kW.

Geothermal systems can also utilise an aquifer, a lake, sea water or any other large body of stationary or flowing water. These are more cost effective since no drilling is required.

Case Study: Water Police, Sydney installs a geothermal cooling system [9]

The Water Police in Sydney is one of the oldest police services in Australia – selected for its new headquarters at Cameron's Cove (on Sydney's famous harbour) a geothermal cooling system for its two-storey 1500 m² (16 146 ft²) headquarters.

Similar to HCTs, the capital costs for geothermal systems are initially more expensive than conventional cooling systems due to the high cost of drilling. However, operational costs are lower because they save water, energy and discharge no emissions to the environment. Energy savings are due to the fact that no external energy source is required for cooling. The compressors in the individual heat pump units of a geothermal unit operate much more efficiently than those in air-source units because the geothermal source/sink temperature is far more stable than that outdoor air and has much less severe high and low extremes.

References

[1]   Sustainable Energy Development Authority. *Energy Savings Manual – Your Profitable Business Strategy*. NSW, Sydney. 2000.

[2]   Galitsky C., Chang S.C., Worrell E. and Masanet E. *Energy Efficiency Improvement and Cost Savings Opportunities for the Pharmaceutical Industry*. Ernest Orlando Lawrence Berkley National Laboratory. www.epa.gov. September 2005.

[3]   Konopacki S., Gartland L., and Akbari H., L Rainer *Demonstration of Energy Savings of Cool Roofs*. www.eed.lbl.gov/EA/Reports/40673.

[4]   Pugh M.D. *Benefits of Water Cooled Systems vs Air–cooled systems for Air-conditioning applications*. Cooling Tower Institute. www.cti.org.

[5] United States Environmental Protection Agency. *Wise Rules for Industrial Energy Efficiency*. www.epa.gov. September 2003.

[6] Sydney Water. *Best Practice Guidelines for Cooling Towers in Commercial Buildings*. March 2004.

[7] Joe A. and Michael G. *Economics of Selecting Air Cooled versus Water Cooled Refrigeration Equipment*. AIRAH. March 2002.

[8] Sydney Water. *The Conserver*. Issue 10. May 2006.

[9] Australian Institute of Refrigeration and Air conditioning. *Harbour cools the Sydney Water Police*. EcoLibrium™. August 2005.

Chapter 7

Steam Systems

7.1 Introduction

Steam systems are vital to commercial, industrial and thermal power plants. In the United States alone, it is estimated that there are over 163 000 boilers in the commercial and industrial sector [1].

The five major steam intensive industries are:

1. food processing
2. pulp and paper
3. chemicals
4. petroleum refining
5. primary metals.

In the U.S. manufacturing sector, these industries account for 71% of the boiler units and 82% of capacity [1]. For instance, in the pulp and paper industry 84% of energy is consumed as steam. Out of which 24% is used in pulping, 20% is used for bleaching and 41% is used for papermaking.

In terms of boiler size, aside from thermal power plants, the chemicals industry, the pulp, paper and oil-refining industries have on average larger boilers than other industries, while not having the most number of boilers. The food industry has the smallest-sized boilers in the industries under discussion. In the commercial property sector, steam systems are also required for heating and refrigeration. However (though numerous), these are (in comparison to the industrial boilers) significantly smaller.

Given the large dependency on steam, in an environment of increasing energy costs and a focus on climate change, increasing steam efficiency allows businesses to be competitive and an opportunity to save energy, water and reduce greenhouse gas emissions.

It is estimated that steam efficiency improvements can reduce a fuel bill by 20–40% [2, 3].

Case Study: What impact does a 30% improvement in steam efficiency has in the US manufacturing sector?

If all US manufacturers improved the efficiency of their steam systems by 30%, they would save approximately 2954 PJ* (2.8 Quadrillion** Btu) of steam energy. This equates to 101 million metric tons of coal, enough to supply the total energy needs of the State of Michigan for a year and reduce CO_2 emissions by 60 million metric tons – and nitrous oxide by 30 thousand metric tons.

Adapted from: U.S. Department of Energy, *Steam Digest* 2000.

* 1 PJ = Peta Joule = 10^{15}J. 29 PJ = 1 million metric tons of coal
** 1 Quadrillion = 10^{15} Btu = 10^{18}J

The following discussion shows how water conservation can achieve savings in steam systems.

7.2   Steam System Principles

The three forms of energy used in industrial processes are as follows.

1.   electricity
2.   direct-fired heat
3.   steam.

Steam is used in a variety of ways, such as

- for process heating and steam tracing
- for pressure control
- for driving mechanical equipment (such as pumps, fans and turbines, and absorption refrigeration)
- as a source of water and heat for process reactions
- for steam jet ejectors to produce a vacuum and component separation (such as in oil distillation)
- to generate electricity.

The wide use of steam is due to its advantages, such as

- low toxicity
- ease of transportability
- high efficiency
- high heat capacity
- low cost relative to other alternatives.

Heat can be stored as sensible heat and latent heat. The advantage of steam is that (unlike other fluids) the majority of the heat is stored as latent heat. Therefore it can be transferred at constant temperature.

As shown in Figure 7.1 a steam system consists of five sections namely

1. · pre-treatment
2. steam generation
3. steam distribution
4. end use and
5. steam recovery.

7.2.1   Pre-treatment

As described in Chapter 2, natural waters contain many impurities that are undesirable for steam generation.

These are

- suspended solids
- dissolved solids
- dissolved gases.

Town water normally does not contain suspended matter but contains varying degrees of dissolved solids such as calcium, magnesium, iron, bicarbonates, carbonates and silica, among others.

The presence of calcium, magnesium, silica and iron contributes to scale formation inside the boiler. The thermal conductivity of scale is an order of magnitude less than the corresponding value for bare steel. Even thin layers of scale as seen from Figure 7.2 reduces heat transfer and if left untreated can result in boiler tube failure. Dissolved gases such as oxygen react and corrode boiler internals. Carbonates break down to carbon dioxide and corrode condensate systems.

Therefore, it is necessary to remove these impurities. Pre-treatment systems predominantly consist of filtration, ion exchange and, lately, reverse osmosis systems. Filtration only removes suspended solids. Ion exchange and reverse osmosis systems remove dissolved solids.

There are three types of ion exchange processes. These are

1. water softening
2. dealkalisation
3. demineralisation.

Water softening is the simplest ion exchange process. The softening of water is readily accomplished with a cation exchanger, usually the sodium form of a strongly acidic, sulphonic acid resin that exchanges sodium ions to calcium and magnesium ions.

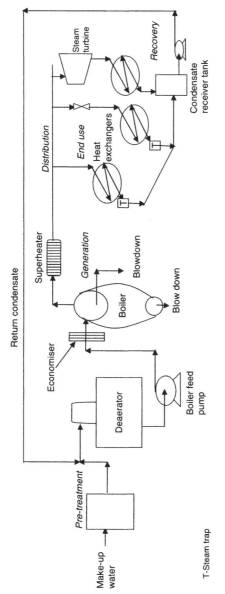

Figure 7.1   Schematic of a typical steam system in an industrial plant

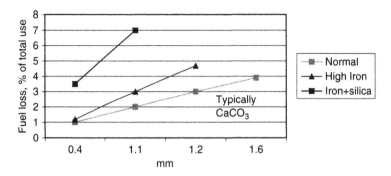

Figure 7.2   Energy loss due to scale deposits

Adapted from U.S. Department of Energy. Industrial Technologies Program, Energy Efficiency and Renewable Energy. *Clean Boiler Waterside Heat Transfer Surfaces.* January 2006.

As shown below:

$$R–SO_3{}^-Na_2 + Ca^{2+} \longrightarrow RSO_3Ca + 2Na^+ \qquad (7.1)$$

$$R–SO_3{}^-Na_2 + Mg^{2+} \longrightarrow RSO_3Mg + 2Na^+ \qquad (7.2)$$

Since the sodium ions replace the calcium and magnesium ions, there is no reduction in total dissolved solids levels. When the resin is exhausted hardness leakage increases and then the resin needs to be regenerated with a saturated brine solution (NaCl) to bring it back to its original capacity. Water softening is appropriate for low pressure boilers normally less than 4020 kPa (600 psi).

$$RSO_3Ca + 2NaCl \longrightarrow R–SO_3{}^-Na_2 + CaCl_2 \qquad (7.3)$$

Dealkalisation refers to the removal of bicarbonate and carbonate alkalinity together with hardness.

For high-pressure systems above 4020 kPa (600 psi) removal of all dissolved solids is essential. Demineralisation removes all dissolved ions (both cations and anions) and replaces them with hydrogen (H^+) or hydroxyl (OH^-) ions. These units are regenerated with sulphuric or hydrochloric acid and sodium hydroxide. However, a problem with demineralisation is the handling and disposal of spent acids and caustic.

Reverse osmosis (RO) units are used to reduce the burden on ion exchange equipment by removing the majority of dissolved solids. The demineralisation unit then acts as a polisher. Chapter 8 discusses membrane technologies.

Other pre-treatment technologies are cold lime softening and hot process softening. In lime softening, temporary hardness represented by Ca and Mg bicarbonate is precipitated as the insoluble carbonate and hydroxide forms respectively.

Electrodialysis (ED) is the passage of an electric current to drive positively and negatively charged ions towards the cathode and the anode respectively through a semipermeable membrane.

Table 7.1 Pre-treatment techniques for common impurities

Substance	Composition	Treatment method
Alkalinity	HCO_3^- and CO_3^{2-}	Chloride anion exchange, demineralisation, lime-soda softening, Reverse Osmosis (RO), electrodialysis (ED) and chemical neutralisation
Ammonia	NH_3	Hydrogen cycle cation exchange, deaeration and chlorination
Carbon dioxide	CO_2	Deaeration, aeration, chemical neutralisation and membrane contactors
Chloride	Cl^-	Demineralisation, RO and ED
Conductivity	Dissolved solids	Demineralisation, RO and Nanofiltration (NF partial demineralisation), ED and vacuum distillation
Fluoride	F^-	Alum coagulation and anion exchange and RO
Free mineral acidity	H_2SO_4, HCl and HNO_3	Anion exchange, chemical neutralisation, RO
Hardness	Ca^{2+} and Mg^{2+}	Sodium and demineralisation, NF and RO, EDI, lime-soda softening
Hydrogen sulphide	H_2S	Aeration and anion exchange using strong base resins
Iron	Fe^{2+} and Fe^{3+}	Aeration, cation exchange, reverse osmosis, chemical precipitation
Manganese	Mn^{2+}	Aeration, cation exchange, reverse osmosis, chemical precipitation
Nitrate	NO_3^-	Demineralisation, biological treatment, RO and ED
Organics	Various naturally occurring organic acids determined as Total Organic Carbon	Anion exchange, granular activated carbon (GAC), UF, NF and RO
Oxygen	O_2	Deaeration, membrane contactors and chemical treatment
Silica	Dissolved and colloidal silica	Colloidal silica – Hot process softening and Ultrafiltration (UF). Dissolved silica – Strongly basic anion exchange, RO
Sulphate	SO_4^{2-}	Anion exchange and RO
Suspended solids		Filtration , microfiltration (MF), UF and NF

Table 7.1 gives a summary of pre-treatment methods for common impurities and Table 7.2 shows the impact of pre-treatment techniques on water-quality parameters.

Dissolved gases such as oxygen and carbon dioxide are corrosive to carbon steel – and ammonia is corrosive to copper alloys. These effects are exacerbated at elevated temperatures. Therefore, these gases and other non-condensable gases are removed to the lowest practical level in deaerators, deaerating heaters and in membrane contactors.

Table 7.2 Comparison of effluent quality from commonly used pre-treatment systems

Technology	Suspended solids	Alkalinity	Hardness	Dissolved silica	Total dissolved solids
Filtration	Reduction close to 100%	No change	No change	No change	No change
Sodium – cycle cation exchange	Prefiltration removes 100%	No change	0–2 ppm hardness in effluent	No change	No change
Two bed strong base/ strong acid Demineralisation	Prefiltration removes 100%	100% removal	100% removal	Reduction of 80–90% removal	100% removal
Reverse Osmosis	Removes 100%	At least 90% reduction	At least 90% reduction	Reduction of 90%	Reduction of 90%

In deaerators, low pressure steam is used to heat the boiler feedwater to its saturation temperature, corresponding to the pressure in the deaerator and in the process oxygen is removed. The feedwater oxygen can be reduced to as low as 40–7 µg/L. Following mechanical deaeration, a chemical oxygen scavenger such as sodium sulphite, hydrazine or carbohydrazide (a safer alternative to hydrazine) is used to remove traces of oxygen. The return condensate also joins the deaerator.

Figure 7.3 shows a photograph of a deaerator.

Figure 7.3   A deaerator in an industrial plant

Courtesy of RCR EA Steel Energy Systems Pty Ltd.

7.2.2 Steam Generation

From the deaerator the boiler feed water in water tube boilers is pumped through the economiser to the steam drum. The economiser preheats the incoming water further with the exiting flue gases.

Boilers are characterised by

- their application
- configuration
- size
- quality of steam produced.

For example, by application, boilers are classified as

- hot water boilers
- steam boilers
- power boilers.

By configuration boilers can be classified as

- firetube
- watertube
- waste heat boilers
- electric boilers
- hot water boilers
- once-through boilers.

In industry, the most common are firetube, watertube and wasteheat boilers. The average capacity of these are about 10 MW (36 000 MMBtu/hr) [1]. Electric boilers are found in commercial applications such as in hotels and have a capacity less than 300 kW. Hot water boilers are designed to heat water to about 121°C and are mainly used in hotels, schools and hospitals. Once-through boilers are found mainly in thermal power plants.

The focus of this chapter is on fire tube, water tube and wasteheat boilers.

7.2.2.1 Firetube Boilers

Firetube boilers are used principally as heating systems for industrial process steam or as portable steam-generating units. They normally produce 230–68 000 kg/hr (500–150 000 lbs/hr). Typical operating pressures for fire tube boilers are below 1014 kPa (150 psig) [4]. These types of boilers are used in applications where steam demands are relatively small or require only saturated steam. In a fire tube there is only one cylindrical steam drum, which serves to store the water and steam. The hot combustion gases pass inside boiler tubes that are contained within the drum. Heat is transferred to water from the hot flue gases in the fire tubes.

Figure 7.4   Photo of a firetube boiler

Courtesy of RCR EASteel Energy Systems Pty Ltd.

Figure 7.4 shows a photograph of a firetube boiler. Firetube boilers use gas or fuel oil in their burners. Firetube boilers are easy to operate and are low in cost to run. Given the large body of water in the drum, they can respond to sudden increases in steam demand.

7.2.2.2   Watertube Boilers

Watertube boilers, due to their ability to withstand high pressures (as much as 20.3 MPa (3000 psi) or higher), are used in high-pressure systems. In watertube boilers, the boiler water is circulated through the tubes while the exhaust gases are outside passing over the tube surfaces. Heat transfer occurs due to radiation heat as well as conduction and convection from the hot combustion gases.

The circulation in water tube boilers can be forced or natural. Natural circulation boilers operate on the principle of differences in density between cold water (higher density) and a steam/water mixture (lower density). Their capacity may range from as little as 3 MW (10 MMBtu/hr) to 3000 MW (10 000 MMBtu/hr). The larger units are found in chemical, oil refining and steel industries. Water tube boilers can consist of factory fabricated package boilers that use oil or gas as fuel to field erected coal fired boilers which can be several storeys high.

While the initial cost of a watertube will be higher than an equivalent firetube boiler, watertube boilers are capable of high efficiencies and can generate either saturated or superheated steam. The superheater further increases the temperature of the steam. These are particularly suited for high-energy high-pressure systems requiring dry steam – such as in steam turbine

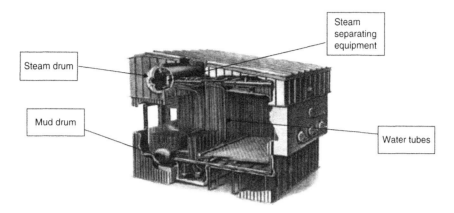

Figure 7.5   A cutaway view of a typical watertube boiler

applications for power generation. Figure 7.5 shows a cutaway view of a typical water tube boiler.

A watertube boiler consists of

- the steam drum
- water tubes
- a mud drum.

The steam drum is the receptacle for steam/water mixture and it separates the steam from the water. The boiler feedwater and chemical dosing line enters the steam drum. The continuous blowdown lines also exits the steam drum. The mud drum (as the name suggests) separates the suspended solids from the water. A bottom blowdown is also located here. The steam drum is bled to control the dissolved solids while the mud drum is bled to remove particulates and sludge. Bottom blowdowns are periodically carried out at predetermined frequencies.

The steam drum continuous blowdown is carried out in accordance with the water-treatment requirements – as dictated by guidelines – such as

- American Society of Mechanical Engineers (ASME) [5]
- British Standard BS 2486 [6]
- Pulp and Paper Institute Guidelines (TAPPI)
- Boiler manufacturers guidelines, such as ABB.

The blowdown water contains useful heat and is therefore a good source of energy to preheat boiler feed water and flash tanks can capture the low-pressure flash steam.

Table 7.3 is a typical guideline for watertube fired boilers.

Table 7.3 Typical guidelines for Industrial Fired Watertube Boilers

Drum operating Pressure							
MPa	0–2.03	2.04–3.04	3.05–4.06	4.06–5.07	5.07–6.09	6.09–6.76	6.78–10.14
Bar	0–21	21–31	31–41	41–52	52–62	62–69	69–103
psig	0–300	301–450	451–600	601–750	751–900	901–1000	1001–1500
Feedwater							
pH	8.5–9.5	8.5–9.5	8.5–9.5	8.5–9.5	8.5–9.5	8.8–9.6	8.8–9.6
Total hardness as mg/L $CaCO_3$	1.0–0.5	0.5–0.3	<0.2	<0.1	<0.1	ND	ND
Dissolved oxygen as mg/L O_2	0.02	0.02	0.007	0.005	0.005	0.005	0.005
Oily substances as mg/L	1–0.5	1–0.5	<0.5	<0.5	<0.2	0.2–0.05	0.2–0.05
Total iron max as Fe mg/L	0.1–0.05	<0.05	0.05–0.03	0.03–0.02	0.02	<0.02	0.02–0.01
Total copper max as Cu mg/L	0.05–0.03	<0.03	0.02	0.02	0.01	0.01	<0.01
Boiler water							
TDS max mg/L	3000	2500–1000	<500	300–200	100–50	50	20
pH	10.5–12	10.5–12	10.5–12	10–11	10–11	9.5–10.5	9.5–10.0
Caustic alkalinity mg/L as $CaCO_3$	50–300	50–150	25–50	10–20	5–10	2–5	2–5
Phosphates as PO_4 mg/L	30–60	30–40	15–20	15–20	15–20	5–10	5–10
Silica as SiO_2 mg/L	<150	<90	<40	30–10	20–10	8–5	5–2

7.2.2.3 Waste Heat Recovery Boilers

Waste heat boilers are unfired boilers and are used in heat recovery applications to generate steam. They may be either watertube or firetube design and use heat that would otherwise have gone to waste.

Ammonia and ethylene plants have unique waste heat recovery boiler designs. For example, in a MW Kellogg ammonia plant, the 101 C heat exchanger is a waste heat boiler which has a *bayonet tube* design. Proper water circulation in these is crucial to minimise tube rupture. Waste heat boilers can sometimes supply close to one-third of the steam demand. They have special requirements, in so much as, being waste heat boilers they may not have sophisticated steam separation equipment and any foaming can decrease steam purity. Heat recovery steam generators (HRSGs) are found in combined cycle cogeneration plants. These are a particular type of waste heat boilers, which generate steam from the exhaust of a gas turbine. Exhaust gases leave the gas turbine at 538° C (1000° F) or higher and can represent more than 75% of the total energy input [7].

7.2.3 Steam Distribution System

The steam distribution system distributes steam from the boiler to the end users such as heat exchangers, steam turbines and process vessels. A well-designed steam distribution system ensures that the steam reaches the end users with a minimum of energy losses. Also once the latent heat has been utilised steam condenses to liquid – known as steam condensate. It needs to be collected and returned to the boiler thus minimising water, chemical and energy wastage.

Consequently, proper performance of the steam distribution system requires careful design practices and effective maintenance. The steam pressure needs to match process requirements. The piping should be the right size and insulated to minimise heat loss. The steam balance between the different steam headers should be carefully balanced. The condensate collection system should have adequate drainage and correctly selected steam traps.

Steam traps are essential for the efficient collection of condensate. Steam condensate is hot pure water. If not removed quickly it hampers heat transfer – since a heat exchanger part full of condensate acts as an insulation barrier reducing the physical size of the exchanger.

Steam condensate also contributes to water hammer, resulting in damage to piping and fittings and loss of heat transfer efficiency. A steam trap is a self-contained valve that automatically drains the condensate from steam-containing enclosure while preventing live steam from passing through. Thus the purpose of the steam trap is to remove condensate, air and CO_2 out of the system as quickly as they accumulate.

There are four types of steam traps:

1. thermostatic
2. mechanical
3. thermodynamic
4. fixed orifice condensate discharge traps (FOCDT).

7.2.3.1   Thermostatic Traps

Thermostatic traps use temperature differential to distinguish between condensate and live steam. This differential is used to open or close a valve. Under normal operating conditions, the condensate must cool below the steam temperature before the valve will open. Common types of thermostatic traps include bellows and bimetallic traps. Thermostatic steam traps are commonly found in low pressure heating systems and are the cheapest type of trap.

7.2.3.2   Mechanical Traps

Mechanical traps use the difference in density between condensate and live steam to produce a change in the position of a float or bucket. The types of mechanical traps include ball float, float and lever, inverted bucket, open bucket and float and thermostatic traps (F&T steam traps). The F&T steam traps operate on both density and temperature principles. The float valve operates on the density principle: A lever connects the ball float to the valve and seat. Once condensate reaches a certain level in the trap, the float rises, opening the orifice and draining condensate. A water seal formed by the condensate prevents live steam loss. Since the discharge is under water, it is not capable of venting air and non-condensables. When the accumulation of air and non-condensable gases causes a significant temperature drop, a thermostatic air vent in the top of the trap opens to vent air (if air is trapped then it will not allow the steam to enter). Shell and tube heat exchangers require fast draining of condensate and F&T steam traps are ideal for this application. Figure 7.6 shows a cut away of a F&T steam trap.

7.2.3.3   Thermodynamic Traps

These use the difference in velocity between condensate and live steam to operate a valve. The disc trap is the most common type belonging to this group. The disc steam trap is time-delayed device that has only one moving part, the disc itself. As long as cold condensate is flowing, the disc will remain open. As soon as steam reaches the inlet orifice, velocity of flow increases, it pulls the disc towards the seat thus closing the disc. It is ideal in situations where space is limited, simple and small. It completely discharges all condensate when open and during intermittent operation for a steady purging action. These are not meant for high-capacity condensate discharge unlike F&T. Figure 7.7 shows a cutaway view of a thermodynamic disc steam trap.

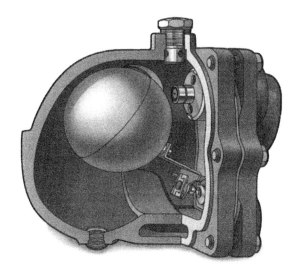

Figure 7.6 A cutaway view of an F&T steam trap

Courtesy of Armstrong International Inc.

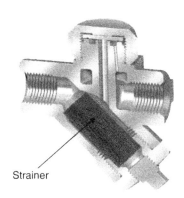

Strainer

Figure 7.7 Cutaway view of a thermodynamic disc steam trap

Courtesy of Armstrong International Inc.

7.2.3.4 *Fixed Orifice Condensate Discharge Traps (FOCDT)*

Fixed orifice condensate discharge traps have no moving parts and in principle therefore requires little maintenance. They have a small orifice that allows for the condensate to pass through while preventing the passage of steam. Unlike conventional steam traps these do not reduce the system pressure. They also give a more consistent heat transfer rate. Moreover, when the orifice is blocked it prevents condensate to pass through, which can be easily detected since the line temperature drops. The FOCDT has an advantage over conventional traps by being more reliable reduces maintenance costs which in turn results in reduced energy consumption.

Case Study

South Manchester University Hospitals Trust is one of the largest patient care providers in the UK. The Trust replaced 86 traps with FOCDT in their laundry at Withington hospital. The laundry has an annual throughput of 1.8 million kg of linen.

Low priority for maintenance of the steam system led to the hospital trust using an additional 500 kg/hr. The replacement with FOCDT led to the hospital saving 19.2% of steam used in the laundry. Reductions in CO_2 and NOx equated to 19% and cost savings of £9,600 per year were realised.

Source: Energy Efficiency Best Practice Programme, *Energy and Cost Savings through the installation of low maintenance steam traps*, Case Study 120. Watford, UK. 2000.

7.3 Steam and Energy Conservation Opportunities

From the previous discussion it is evident that there are a number of steam and energy conservation opportunities in steam plants. These are shown in Table 7.4.

Some of these measures are described below.

Table 7.4 Steam conservation opportunities

Area	Action
Maintenance	Repair steam leaks
	Maximise condensate recovery
	Install continuous blowdown heat recovery
Operating practices	Minimise vented steam
	Reduce steam system operating pressure
	Improve blowdown practices
	Isolate steam from unused lines
	Reduce excess boilers on standby or install smaller boiler
	Use high-pressure condensate to make low-pressure steam
	Reduce deaerator vent steam rate
Water treatment	Maintain clean boiler heat transfer surfaces
	Minimise blowdown
	Reduce boiler water dissolved solids
	Improve condensate recovery by improving condensate chemistry and preventative maintenance of steam traps

7.3.1   Repair Steam Leaks

Steam leaks are often found at

- valve stems
- unions
- pressure regulators
- equipment connection flanges
- pipe joints.

While saturated steam is visible to the naked eye, high-pressure superheated steam is not and poses a safety risk. A typical saving of 1.4% fuel saving is realisable. Table 7.5 shows the steam losses as a function of hole diameter and pressure.

Table 7.6 shows the annual cost of steam leaks from a 1035 kPa (150 psig) steam pressure line as a function of hole diameter.

Table 7.5   Steam loss kg/hr as a function of pressure and hole diameter

Hole diameter (mm)	Hole diameter (in.)	Steam Loss kg/hr			
		Steam pressure kPa (psi)			
		104 (15)	690 (100)	1035 (150)	2070 (300)
0.8	1/32	0.4	1	2	–
1.6	3/16	2	6	9	16
3.2	1/16	6	24	34	65
4.5	1/4	14	54	77	147
6.5	1/8	25	95	136	261
9.5	3/8	55	214	307	586

Adapted from US EPA Steam Challenge.

Table 7.6   Cost of steam leaks from 1035 kPa steam pressure

Size of orifice (mm)	Size of orifice (in.)	kg of steam wasted per year	Total cost per year (A$/yr)*
0.8	1/32	19,008	171
1.6	1/16	74,304	676
3.2	1/8	297,216	2,710
4.8	3/16	666,144	6,078
6.4	1/4	1,187,136	10,833
9.5	3/8	2,672,352	24,384

* Steam cost at A$9/ton.

7.3.2 Maximise Condensate Recovery

Optimising condensate return for reuse as boiler feedwater is a profitable means of reducing fuel costs and water usage while increasing boiler system efficiency.

Strategies to recover condensate are to

- maximise condensate return
- reduce steam leaks
- reduce venting of steam
- isolate steam from unused lines
- monitor condensate quality
- minimise waterlogging of pipes.

If steam traps have not been maintained for 3–5 years it is estimated on average 15–30% of steam traps may have failed. Regularly scheduled maintenance should reduce this to fewer than 5% of traps [2]. The dollar impact of poor trap maintenance can be seen from the example cited in Table 7.7.

The industry standard is to carryout inspections of steam traps as follows:

- high pressure 1034 kPa (150 psig) – weekly to monthly
- medium pressure – monthly to quarterly
- low pressure – annually.

Table 7.7 shows the potential water loss from a system that has 1000 steam traps.

As more condensate is returned, less make-up water is required, saving on both water and pre-treatment costs. The high purity of condensate allows for greater boiler cycles of concentration, thus reducing water and energy losses to blowdown. Since condensate is typically around 80°C (176°F) the added benefit of returning hot condensate translates to reduced heated cost of make-up water. Additional savings can also be made in reduced water-treatment chemicals and sewer discharge costs.

Table 7.7 Estimation of daily water loss from steam traps

Total number of steam traps	1000
Average steam trap failure rate	20%
Minimum leakage rate from a failed steam trap	10 kg/hr
Estimated number of failed steam traps	$1000 \times 0.2 = 200$
Loss of live steam	$200 \times 10/1000 = 2$ tons per hour
Percentage flash off	10%
Daily water loss, m³/d	$= (1 - 0.1) \times 2 \times 24 = 43.2$
Annual cost*	$= 2 \times 24 \times 365 = \$157,680$

* Steam cost at A$9/ton

A simple calculation indicates that energy in the condensate can be more than 10% of the total steam energy content of a typical system.

Water and energy savings from increased steam condensate recovery

Assume that the temperature of condensate is 80°C (176°F) and that the temperature of make-up water is at 15°C (59°F). The steam pressure is 700 kPa. The percentage of heat contained in the condensate can be calculated as:

h_c = enthalpy of condensate at 80°C (176°F) $\qquad$ = 334.9 kJ/kg

h_m = enthalpy of make-up water at 15°C (59°F) $\qquad$ = 62.9 kJ/kg

h_s = enthalpy of steam at 800 kPa absolute (114 psia) = 2768 kJ/kg

Heat remaining in condensate $= (h_c - h_m)/(h_s - h_m)\%$

$$= (334.9 - 62.0)/(2768 - 62.9) \times 100$$

$$= 10.1\%$$

Worked example

Assume that the steam system returned an additional 2 ton/hr of condensate at 80°C. The plant operates for 8000 hrs annually with an average boiler efficiency of 82%. The temperature of make-up water is 15°C.

The water usage charges is $1.00/m^3$ and the sewer usage charge is $1.00/m^3$. The cost of water treatment is $2.00/m^3$. The gas cost is $9.00/GJ$. Assume a 10% flash steam loss. Calculate the overall annual savings.

Annual water, sewage and chemical savings = (1 – flash steam traction) × (condensate load in tons/hr) × Annual operating hours × (total water costs $/m^3$)

$$= (1 - 0.1) \times 2 \text{ tons/hr} \times 8000 \text{ h} \times \$4/m^3 = \$57\,600/\text{yr}.$$

Annual Fuel Savings = (1 – Flash steam fraction) × Condensate load tons/hr × Operating hours × Make-up water temperature rise × Gas costs in $/GJ / Boiler efficiency

$$= (1 - 0.1) \times 2 \text{ tons/hr} \times 8000 \text{ h} \times (80°C - 15°C) \times \$9/GJ/0.82 = \$10\,273$$

Total Annual Savings due to return of an additional 2 tons/hr of condensate = $57\,600 + 10\,273 = \$67\,873$.

7.3.2.1   Condensate Quality and System Protection

Corrosion in condensate systems can limit the quality and quantity of returned condensate because, dissolved and insoluble iron and copper, oil and grease can deposit on boiler heat transfer surfaces. This reduces heat transfer efficiency and could cause premature tube failure.

Dissolved gases in condensate such as CO_2, Oxygen (O_2) and ammonia (NH_3) contribute to general (and pitting) corrosion of process equipment, lines and tanks. The major source of CO_2 is the breakdown of feedwater bicarbonate and carbonate alkalinity in the boiler. In water, CO_2 forms carbonic acid and since condensate is extremely pure even small quantities of carbonic acid can significantly depress the pH to 4.5–5.5. A classic indication of carbonic acid attack is shown in the photo on the left of Figure 7.8. The higher the temperature the more aggressive the acid is to ferrous metals. In the presence of O_2 the resulting corrosion rate can accelerate 10–40 times faster than the rate of either gas alone [8]. Oxygen causes pitting attack and this is shown in Figure 7.8 photo on the right. NH_3 is corrosive to copper and copper alloys. The corrosion by-products get carried to the boiler and deposit in boiler tubes. This in turn reduces heat transfer and the usual remedy is to dump the condensate.

A better way is to reduce corrosion by the addition of chemical inhibitors or using a magnetic filter. Condensate corrosion inhibitors known as neutralising amines react with the CO_2 and elevate the pH to 8.8–9.2. The most common is amine-based organic compounds such as morpholine and cyclohexylamine. Filming amines also protect condensate systems by forming a film and preventing the oxygen to attack the pipe interior. Magnetic filters remove particulate iron in condensate streams. They typically remove 95% of magnetite present [9]. In food plants due to food safety considerations, only a very limited number of neutralising amines can be used, magnetic filters are commonly used to remove the iron in steam condensate.

Condensate can get contaminated from process fluids. When this happens, the condensate needs to be segregated from the clean condensate. Contaminants commonly found in condensate range from sugars, hardness

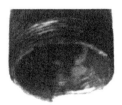

Carbonic acid attack showing   Pitting corrosion due to
the metal thinning.   oxygen attack.

Figure 7.8   Corrosion of condensate due to carbonic acid and oxygen

Courtesy of GE Water and Process Technologies.

leakage, fibrous matter, animal fats, grease, naphtha, kerosene, hexanes and fatty acids, among others.

The most common is oil contamination. If left untreated the oily condensate contributes to foaming and sludge deposits in the boiler. Therefore, it is common practice to dump this source of water. The most immediate step is to eliminate the contamination by fixing leaks. If the contamination is severe, then dumping the condensate may be the short-term answer. However, there are technologies to effectively treat oily condensate. These technologies are described in Chapter 8.

7.3.2.2 Minimise Water Logging of Pipes

Incorrect location of the steam trap can cause water logging of pipes and result in loss of output, water hammer and erosion of pipes. When two or more vessels are connected to the same pipeline and share a common steam trap known as group trapping, it can lead to water hammer.

It is important to ensure that the pipes are sloped correctly to minimise water hammer. The main should have a slope of 12.5 mm (1/2 in.) in 3 m (10 ft) in the direction of the steam flow. Also ensure there are sufficient steam traps in risers and along horizontal lines and unused steam lines need to be isolated.

Case Study

A large specialty paper plant reduced its boiler make-up water rate from about 35% of steam production to between 14 and 20% by returning additional condensate. Annual savings added up to more than US$300 000.

Source: US Department of Energy. Steam Tip Sheet Number 8. *Return condensate to the boiler*. June 2001.

7.3.3 Minimising Boiler Water Blowdown

7.3.3.1 Blowdown Control

Proper control of blowdown is a critical part of boiler operation. While insufficient blowdown may lead to deposits and carryover, excessive blowdown will waste water, heat and chemicals. All boiler feedwaters contain some solid impurities. When the feedwater is evaporated in a boiler, steam is formed from the water, leaving the solids behind. As a result, the solids gradually build up in the boiler water. To control these solids, boiler water is removed (i.e. blown down) and replaced with lower solids feedwater.

To keep the solids from building to unacceptable levels, the amount of solids removed by blowdown must equal the amount brought in with feedwater as shown in Figure 7.9. The quantity of blowdown depends on both the amount of feedwater solids and the level of solids that can be tolerated in

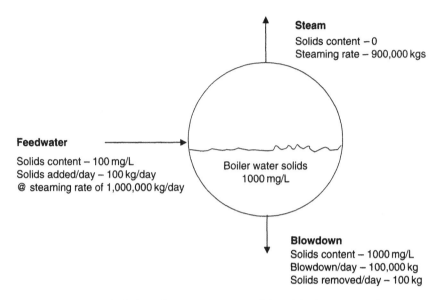

Figure 7.9   The balance between incoming solids with boiler concentrations

a particular boiler – based on the pressure of the boiler as per the guidelines recommended by manufacturers or national standards (Table 7.3).

Minimising boiler water blowdown will reduce water, chemicals and energy costs. On the other hand, increasing solids in the boiler water can cause steam contamination.

The optimum blowdown rate is a function of

- boiler type
- boiler pressure
- water treatment chemistry
- make-up water quality.

Blowdown rates for softened water can be around 5% and for demineralised makeup – about 1%

Blowdown percentages is calculated as

$$\% \text{ Blowdown} = \frac{\text{volume of blowdown water}}{\text{volume of feedwater}} \times 100 \qquad (7.4)$$

or

$$\% \text{ Blowdown} = \frac{\text{conductivity of feedwater}}{\text{conductivity in boiler water}} \times 100 \qquad (7.5)$$

The second equation shows that per cent blowdown increases as (a) feedwater solids increases and (b) boiler water solids decrease.

There are four basic ways to minimise boiler blowdown. These are

1.   improve boiler blowdown control
2.   alter treatment chemicals to minimise feedwater conductivity
3.   improve boiler feedwater quality.

Examining each alternative enables an assessment of potential savings.

1. Improve boiler blowdown control and install automatic blowdown control

The best way to illustrate this is to use an example. The following parameters will apply.

The boiler water conductivity ($Cond_{BW}$) control limit is	600–1000 $\mu S/cm$
Average $Cond_{BW}$ in boiler water is	700 $\mu S/cm$
Boiler Feedwater conductivity ($Cond_{BFW}$) is	50 $\mu S/cm$
Steaming Rate	50 tons/hr
Annual hours of operation	8000
Cost of gas	$6/GJ

Concentration ratio $(CR) = Cond_{BW}/Cond_{BFW} = 700/50 = 14$

Based on the conductivity limits of 600 and 1000 $\mu S/cm$, the CR can vary from 12 to 20.

Boiler make up = Steam rate $\times CR/(CR - 1) = 50 \times 14/(14 - 1) = 53.85$ tons/hr

Blowdown rate = Make up − steam rate = $53.85 - 50 = 3.85$ tons/hr

If the average $Cond_{BW}$ is increased to 900 $\mu S/cm$, then the new CR is $CR_1 = 900/50 = 18$

New blowdown rate = $50 \times 18/(18 - 1) = 52.94 - 50 = 2.94$ tons/hr

Reduction in blowdown rate = $3.85 - 2.94 = 0.91$ tons/hr

If cost of water is $1.0/m^3$ and sewage charges are $1.0/m^3$ and chemical treatment costs are $3.0/m^3$, then the annual water and chemical savings = $0.91 \times 8000 \times 5.0/m^3 = 36400/yr.$

To achieve these savings strict control of boiler blowdown is required. An automatic blowdown controller will pay for itself in one year.

2. Alter the boiler treatment programme to minimise boiler blowdown

Boiler water treatment chemicals such as phosphate and sulphite add dissolved solids to the boiler. By altering the treatment programme where there is a reduced contribution of dissolved solids from the chemicals the boiler water blowdown can be minimised.

For example, sodium sulphite is commonly added to mitigate against oxygen-induced corrosion in the pre-boiler and after-boiler sections. However, sodium sulphite adds solids to the boiler water. By converting the treatment programme to a volatile chemical such as hydrazine or its derivatives such as carbohydazide, this is avoided.

3. Improve boiler feedwater quality

As discussed, softeners do not reduce the dissolved solids. By converting to a demineralisation system, the dissolved solids are removed completely and the boiler water cycles increase significantly reducing boiler blowdown.

For instance, using the previous example, let

Boiler water conductivity ($Cond_{BW}$)	$700 \mu S/cm$
Boiler Feedwater conductivity ($Cond_{BFW}$)	$50 \mu S/cm$
CR	14

If $Cond_{BFW}$ is decreased to $20 \mu S/cm$, then, new $CR = 700/20 = 35$

If the blowdown is continuous, heat recovery can result in fuel savings.

7.4 Calculating the "True" Cost of Steam

Calculating the true steam costs is essential for water and energy conservation projects to correctly reflect the economic opportunities. Incorrect utility costing can lead to poor investment decisions. Good projects can be discarded and bad projects implemented.

To avoid such mistakes plant managers need to use appropriate steam pricing methods, taking into account all the parameters that impact energy costs which includes fuel, condensate, power generation, cooling water, water treatment, labour, maintenance and water and blowdown discharge costs to the sewer.

It is common to use average costs of generation at a particular production rate. The total operating costs which includes all of the above costs are divided by the total amount of steam produced. Whilst this approach produces a convenient benchmark it may not correctly reflect the true picture. This is especially so when there are multiple boilers and steam turbines. Without going into too much detail the cost of generating steam needs to include the following [10]:

Cost of fuel (C_F)
Cost of raw water (C_w)
Cost of boiler feed water treatment (C_{BFW})
Feedwater pumping power (C_p)
Combustion air fan power (C_A)
Trade waste charges for boiler blowdown (C_B)
Ash disposal (for coal fired and bagasse boilers) (C_D)
Environmental emissions control (C_E)
Maintenance materials and labour (C_M)

Cost of steam generation $= C_F + C_W + C_{BFW} + C_p + C_A + C_B + C_D + C_E + C_M$
Fuel costs is by far the major cost of steam.

$$C_F = \frac{\text{Fuel cost} \times (h_S - h_{fw}) \times T}{\text{Boiler efficiency}} \tag{7.6}$$

where

Fuel cost is given in \$/GJ (\$/MM Btu)
h_s – enthalpy of steam kJ/kg (Btu/lb)
h_{fw} – enthalpy of feedwater kJ/kg (Btulb)
T – number of hours of operation per year

Overall boiler efficiency is primarily a function of flue gas temperature and will be in the range of 75–85%.

For oil and gas fired facilities the other costs can be approximated as 30% of the fuel costs and therefore the cost of generation is simplified as

$$C_G = C_F(1 + 0.3) \tag{7.7}$$

The above method gives the average cost of generating steam. However, for water and energy conservation projects it is more useful to use the marginal cost of generation since most of the infrastructure is already paid for and the option to save is the incremental cost of gas or oil.

$$\text{Marginal cost} = \frac{\text{Incremental operating cost}}{\text{Incremental steam consumption}} \tag{7.8}$$

It is not the intention of the book to go into the details of calculating the marginal cost of steam since it depends on a number of variables.

References

[1] Energy and Environmental Analysis Inc. *Characterisation of the US Industrial/Commercial Boiler Population*. Arlington Virginia. May 2005.
[2] Hart F.L. and Jaber D. *Best Practices in Steam System Management*. Steam Digest. US Department of Energy. Washington. 2001.
[3] Jones T. Alliance to Save Energy. *Steam Partnership: Improving Steam Efficiency Through Market Partnerships*. Steam Digest 2000. US Department of Energy. Washington.
[4] Kemmer F. (ed.). *Nalco Water Handbook*. 39.31. McGraw-Hill Inc. New York. 1988.

[5]   American Society of Mechanical Engineers. *Consensus on Operating Practices for the Control of Feedwater and Boiler Water Chemistry in Modern Industrial Boilers.* 1994.

[6]   British Standard BS 2486:1997. *Recommendations for Treatment of Water for Steam Boilers and Water Heaters.* 1997.

[7]   US Department of Energy. Energy Efficiency and Renewable Energy. *Improving Steam System Performance – A Source book for Industry.* Washington D.C. October 2004.

[8]   Nalco Chemical Company. *Condensate system protection.* Technifax TF-150. 1985.

[9]   Bloom D. *Strategies in Optimising Condensate Return.* US Department of Energy, Office of Energy Efficiency and Renewable Energy: Steam Digest Vol. 1V. 2003.

[10] US Department of Energy. *Energy Efficiency and Renewable Energy.* A Best Practices Steam Technical Brief – How to Calculate the True Cost of Steam. Washington DC. September 2003.

Chapter 8

Industrial Water Reuse Technologies

8.1   Introduction

'Water reuse' and 'recycled water' are terms that are often confused with each other. In order to clarify the difference, water reuse as defined in this book is on-site water reuse. The process water or effluent is used within the site – with or without further treatment. Recycled water or reclaimed water, on the other hand, is the water available from a sewage-treatment plant after undergoing secondary, tertiary or advanced treatment. Figure 8.1 shows graphically the two types of systems.

There is sufficient information available on reclaimed water applications. The most common application of reclaimed water use is in power plant and industrial cooling water systems, irrigation of golf courses, parks and agricultural crops. This chapter focuses on industrial water reuse.

Traditionally, industrial wastewater treatment is undertaken to meet regulatory compliance. Consequently, the treatment of wastewater effluent has been seen more as *the cost of doing business* rather than as a valuable resource. Therefore, the minimum treatment required to comply with a regulatory requirement is fulfilled. In recent years this view is changing. There is the realisation that security of water supply, drought, scarcity of water and increased costs for potable water and sewer discharge are forcing some companies to rethink this minimalist strategy. As a result of this new thinking, some organisations have embraced the notion of *zero discharge*. While the notion of zero discharge is appealing, in practice, achieving it is far from simple. High upfront capital costs can make zero discharge unviable except in some rare cases. The end of pipe treatment is always costlier than reuse at the source. For a project to be feasible, the proposed treatment technologies need to be reliable, economical, meet water-quality specifications and be safe to use, as determined by regulatory authorities.

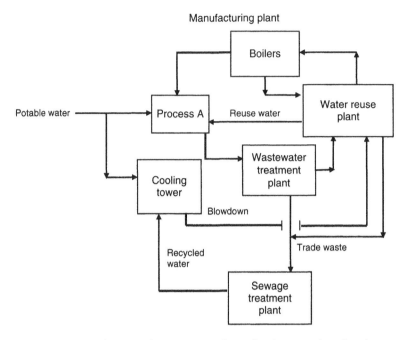

Figure 8.1   A schematic of water reuse and recycling in a manufacturing plant

Two questions always arise when the topic of water reuse comes up. They are

1.   What water quality is acceptable?
2.   What volume of water is available for reuse?

'Acceptable' water quality is dictated by end use. The potential volume of water available for reuse is dictated by the type of industry sector, the achievable cost savings and the prevailing regulatory environment that enables the reuse of water. For instance, in Japan [1] water recovery from industry is significant. The recovery rates fall into three distinct groups. Over 80% of water used is recovered in the steel, chemical and transport machinery sectors.

Table 8.1 shows the water recovery rates in Japan.

In some industries such as in the pulp and paper industry technological advances such as alkaline sizing are enabling greater mill closure with consequent higher recovery levels of water and fibre.

Another way to look at it is how much water is discharged to the sewer from an industry sector? For example, in the food and beverage processing sector 50–89% of the water is discharged to the sewer.

This chapter provides a step-by-step approach to water reuse and examines the current technologies for water reuse applications.

Table 8.1 **Recovery rates for Japanese industries [1]**

Industry	Feed ML/day	Recovered water ML/day	Total consumption ML/day	% Recovery rate
Food	2,586	1,527	4,113	37.1
Drink	810	366	1,175	31.1
Textile and dyeing	2,358	61	3,020	21.9
Pulp and paper	8,935	6,591	15,526	42.5
Chemicals	8,761	40,140	48,901	82.1
Plastics	835	1,757	2,592	67.8
Ceramics	1,223	2,660	3,883	68.5
Iron and Steel	3,917	34,460	38,377	89.8
Non-ferrous metals	891	2,447	3,338	73.3
Metal products	525	478	1,003	47.6
Electrical machinery	1,544	3,763	5,307	70.9
Transport machinery	932	11,290	12,222	92.4

8.2 A Step-by-Step Approach to Water Reuse

In general, water reuse is an integral part of a holistic water management philosophy. The following points need to be considered when conceptualising a water reuse project:

- Follow the principles of resource minimisation hierarchy – *avoid, reduce, reuse and recycle* as shown in Figure 8.2. By avoiding the use of water and reducing the pollutant load, the need for subsequent disposal is eliminated. Reducing water use is cheaper than water reuse.
- Consider source reduction first. End-of-pipe treatment is a more expensive option.

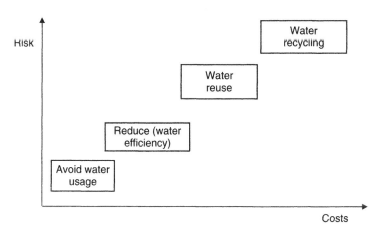

Figure 8.2   Water minimisation hierarchy

- Address the easy options before tackling the more expensive options.
- Segregate the more contaminated streams from the diluted streams and treat them separately.

A step-by-step guide for water reuse projects is summarised below:

1. Establish the goals of the project.
2. Define the project boundaries.
3. Gather data.
4. Identify and evaluate the water reuse projects – relative to goals.
5. Conduct a water reuse technical assessment – desk top as well as a pilot study.
6. Implement the new water-reuse model or design.
7. Commission new system.
8. Monitor operation.
9. Review and update the model or design as needed.

8.2.1   Establishing the Goals of the Study

Clearly specify the goals of the study, the business drivers and alignment with corporate strategy.

8.2.1.1   Goals
Goals of water reuse can be

- meeting EPA or local water authority trade waste regulatory compliance as a condition of operation
- cost savings
- environmental considerations
- good corporate citizenship.

The drivers will dictate the priority and the financial backing at the time of project approval. For instance, if the goal was trade waste compliance then the financial considerations will not be that important, since without achieving trade waste requirements it is difficult to discharge effluent to the sewer.

8.2.1.2   Project Boundaries
What are the boundaries of the study? Is the water reuse study to be undertaken site wide or be limited to a specific area? If the study is undertaken too narrowly then other potential options may be overlooked making the financial justification of the project difficult.
    What is the time frame to carry out the study?

8.2.2   Gather Data

Before embarking on a water reuse project it is essential to gather all relevant information. This includes

- Developing a water balance of the site to identify the water sources and water users.
- Analyse the water streams in order to have a general idea of water quality such as conductivity, suspended solids, oil and grease and other contaminants.
- Understand the production process specific to the plant.
- Locate past plant drawings and any changes made to the plant.
- Understand the water-quality requirements acceptable to the plant. Table 8.2 shows industrial process water-quality requirements [2].
- If there are any health and safety standards that need to be achieved in terms of pathogens such as *Legionella*, *Listeria* and *Campylobacter* then these need to be listed. For instance, the pharmaceutical industry must comply with the *Good Manufacturing Practice code (GMP)*. Similarly, the food industry needs to comply with local and international food regulatory standards and HACCP Guidelines (refer to Chapter 13 for details on HACCP).
- List the trade waste standards applicable to the site.
- List if there are any other regulatory approvals that need to be considered such as local council building approvals and so on.
- Determine the current cost of water, wastewater, trade waste charges for pollutants and what the likelihood might be of these charges increasing.
- What is the typical financial justification hurdle rate applied by the organisation? Typically this varies from 2–5 year payback. Some organisations dictate an internal rate of return (IRR) better than 25% for a project to proceed. For details on IRR, refer to Chapter 9.
- Are plant expansions or contractions likely? Will these plans dictate more or less water use?
- Is any large water using equipment earmarked to be upgraded or become more water efficient?
- Are environmental considerations given a lower hurdle rate? That is, an extended payback period.
- Are there any other considerations that would make it more attractive?

8.2.3   Identify the Project

All personnel who could have knowledge of the water systems need to be invited to a preliminary brainstorming meeting to discuss, identify and prioritise the water reuse options. The facilitator needs to list all

Table 8.2 Typical Industrial Process Water-Quality Requirements [2]

Parameter*	Pulp & Paper			Chemical	Petrochemical and Coal	Textiles		Cement
	Mechanical Piping	Chemical Unbleached	Bleached			Sizing Suspension	Scouring Bleach & Dye	
Copper	–	–	–	–	0.05	0.01	0.1	–
Iron	0.3	1.0	0.1	0.1	1.0	0.3	0.01	2.5
Manganese	0.1	0.5	0.05	0.1	–	0.05	–	0.5
Calcium	–	20	20	68	75	–	–	–
Magnesium	–	12	12	19	30	–	–	–
Chlorides	1,000	200	200	500	300	–	–	250
Bicarbonate	–	–	–128	–	–	–	–	–
Nitrate	–	–	–	5	–	–	–	–
Sulphate	–	–	–	100	–	–	–	250
Silica	–	50	50	50	–	–	–	35
Hardness Total	–	100	100	250	350	25	25	–
Alkalinity Total	–	–	–	125	–	–	–	400
Total Dissolved Solids	–	–	–	1,000	1,000	100	100	600
Total Suspended Solids	–	10	10	5	10	5	5	500
Colour	30	30	10	20	–	5	5	–
pH	6–10	6–10	6–10	6.2–8.3	6–9	–	–	6.5–8.5

* All values in mg/L except colour and pH.

the possible options. Then prioritise the options for further investigation based on

- Water-saving potential
- Cost saving potential
- Technical feasibility
- Technical risk
- Operator complexity
- Time frame – short-term or long-term projects.
- Capital expenditure required
- Return on Investment
- Meeting other organisational drivers.

Carry out a detailed technical review of the selected project(s).
Some examples of water reuse projects include

1. Reuse of wastewater effluent as cooling water make-up – quite common in petrochemical and oil refineries where the cooling water make-up needs are large with stringent wastewater discharge requirements. Challenges are in ensuring that the water is not corrosive to the system and is safe for use.
2. Use of wastewater in boilers – not a common practice but technically feasible. Requires capital expenditure to pre-treat the wastewater to meet boiler feedwater specifications.
3. Reuse of effluent for toilet flushing and washing of floors – challenges are in ensuring that the water is safe for accidental contact through aerosols.
4. Reuse of textile effluent in dye houses – reduces water usage, chemicals and energy usage. Technical challenges are in the removal of oil, colour, dissolved solids, salt and fibre.

Case Study: DPK Australia Pty Ltd – Clean Water Project

The Australian textile industry has been devastated by imports from China. Those left behind are those that have the entrepreneurial vision to compete in niche markets and are very cost conscious. DPK Australia, a family-owned company established in 1981 manufactures and supplies innovative knitted fabrics using Australian Merino wool and other luxury fibres. It partnered with Sydney Water's Every Drop Counts Business Program, to reuse 75% of the wastewater approximately 175 million litres per year that is currently going to waste. In February 2006, DPK was successful in receiving A$525 000 from the NSW government towards a A$1.25 million project to install an innovative vibrating membrane technology for its reuse project.

5. Use of boiler blowdown as cooling water makeup – despite the high heat content, boiler blowdown is a source of good-quality water. However, boiler water polymers can interfere with cooling water treatment chemical programmes.

8.2.4   Technical Assessment

Technical reviews can be conducted as

- desktop technical and financial reviews
- pilot plant trials.

Desktop technical reviews are a screening process to quickly eliminate the non-feasible projects. This assessment can be done at low cost.

Once the options have been identified it is always preferable to carry out an on-site trial or a pilot plant assessment especially when the capital cost is high – to validate the results of the desk top review. Even though pilot plant trials are costly it is only a fraction of the actual plant costs and well worth the expenditure to minimise the risk of equipment non-performance. On-site pilot plant trials give the organisation a higher degree of confidence of potential plant performance and a more realistic assessment of savings and costs. The scope of the trial needs to be developed with the aforementioned needs in mind.

8.2.5   Implementation

Once a firm cost has been received from the contractor, then it is possible to do a sensitivity analysis. If the sensitivity tests show that the project is sensitive to interest rates, capital costs, operational savings, production schedule or some other factor then they need to be examined closely.

Once all of these factors have been examined then a business case needs to be written justifying the project as per the original goals.

A new water balance of the site based on the reuse project can be drawn up.

8.3   Pollutants Found in Reuse Streams

Pollutants found in industrial effluent streams can be segregated as follows:

- inorganic
- organic
- dissolved
- suspended
- gaseous

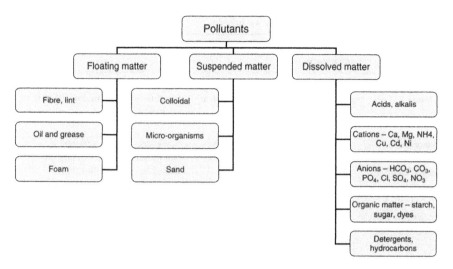

Figure 8.3 Classification of pollutants

- non-ionic
- biological
- pH.

Classification of pollutants is shown in Figure 8.3.

Wastewater quality in industrial effluent is varied. A useful measure of organic loading is the measurement of biological oxygen demand (BOD) and chemical oxygen demand (COD). Refer to Chapter 2 for more details.

The biodegradability of a solution is a measure of the breakdown of chemicals contained in the solution by bacteria. The ratio of BOD/COD determines the theoretical biodegradability of a chemical mixture. If the ratio is greater than 0.4 it is considered to be readily biodegradable. Table 8.3 shows industries where these contaminants are commonly found and their methods of treatment.

8.4 Removal of Pollutants

As Table 8.3 shows, the technologies to remove these pollutants from industrial wastewater streams are diverse and range from

- simple screening, settling and filtration
- chemical and biological treatment
- ion exchange
- reverse osmosis
- electrodialysis and brine concentration.

This section examines these technologies briefly.

Table 8.3 **Wastewater contaminants by industry**

Industry	Characteristics of wastewater	Treatment options
Food processing Dairy	High in dissolved organics – mainly protein, butterfat and lactose. BOD 2,700; COD 4700. BOD/COD = 0.57	Equalisation, dissolved air flotation, aerobic or anaerobic biological treatment
Meat and poultry processing	High in dissolved and suspended organics including protein, blood, greases, fats and manure. BOD 1430. COD 2746. BOD/COD = 0.52–0.83	Screening, gravity separation, neutralisation, DAF, biological treatment, coagulation and precipitation
Vegetable and fruit canneries	High in dissolved and suspended organics from natural products Jams, jellies BOD 2400; COD 4000. BOD/COD = 0.60	Screening, equalisation, gravity separation, neutralisation, biological treatment
Bakery products	BOD 3200, COD 7000 BOD/COD = 0.46	Biological treatment
Iron and Steel	Phenol, tars, free ammonia, cyanide, iron, suspended solids, oil and grease, mill scale, heavy metals	Equalisation, neutralisation, coagulation
Organic chemicals	Dissolved organics, including acids, aldehydes, phenolics and free and emulsified oils	Gravity separation, flotation, equalisation, neutralisation, coagulation, chemical oxidation, biological treatment, adsorption
Petroleum refining	Phenolics, free and emulsified oils and other dissolved organics	Gravity separation, flotation, equalisation, coagulation, chemical oxidation, biological treatment and membrane technology
Pulp and paper	Fibres, dissolved and suspended organics, high BOD and COD	Screening, gravity separation, biological treatment, chemical oxidation and membrane technology
Plastics and resins	Dissolved organics, including acids, aldehydes, phenolics, cellulose, poly vinyl alcohol, surfactants and oils	Gravity separation, flotation, coagulation, chemical oxidation, solvent extraction, adsorption, biological treatment
Textiles	Dissolved and suspended organics, salt, heavy metals, sulphides, dyes, BOD 80–6000 and COD 800–30,000. Total Suspended Solids – 15–8,000; Cl 1,000–1,600; heavy metals; BOD/COD = 0.2–0.54	Equalisation, neutralisation, coagulation, adsorption biological treatment, membrane technology and ozonation

Table 8.3 *(Continued)*

Industry	Characteristics of wastewater	Treatment options
Coke and gas	High in phenolics, benzene, CN, oil and grease, volatile organics, polynuclear aromatic hydrocarbons, Se, Hg, ammonia and other dissolved organics	Equalisation, flotation, adsorption, biological and chemical oxidation, granular activated carbon, solvent extraction
Landfill leachate	High in BOD, COD, Cd, Pb, Zn, Cl, Fe, Ammonia, P BOD/COD – 0.05–0.49	Biological, ozonation, ammonia stripping, ion exchange and membrane treatment

8.4.1   Order of Removal

The order of removal is given below:

- Primary treatment – adjustment of pH and temperature, removal of suspended solids, oils and fats and heavy metals.
  Technologies used in primary treatment consist of screening, sedimentation, hydrocyclones, settling, filtration, dissolved air flotation and centrifugation.
- Secondary treatment – Biodegradable carbon compounds, nutrients (nitrogen and phosphorus).
- Tertiary treatment – Removal of dissolved solids and disinfection to eliminate pathogens.

These treatment methods will be discussed in the ensuing sections.

8.5   Removal of Suspended Solids

Suspended solids in wastewater can be classified as shown in Table 8.4. Knowledge of the *particle diameter and settling time* allows us to select the most appropriate removal technique.

Table 8.4 Classification of suspended solids

Solids greater than 25 mm (1 in.) in diameter. These can obstruct and damage downstream treatment operations.
Solids in the range 10–25 mm in diameter. Examples are sand gravel and are classified as grit.
Settleable solids with diameters between 10^{-3} mm(1 μm) and 10 mm in diameter.
Colloidal substances. Substances in diameter greater than 10^{-6} mm –10^{-3} mm (10^{-3} μm to 1 μm).

8.5.1 Screening

Screening is the simplest way to remove suspended matter and oversized material which might damage equipment or disrupt the treatment process. Basically, it involves placing a screen or plate with fixed openings in the path of the wastewater flow. All suspended solids larger than these openings are trapped on the upstream side and removed by mechanical cleaning devices. The opening size required for a specific application is governed by the purpose of the screen, the downstream unit operation and the particle size that the screen can effectively remove.

The screens can be classified as follows:

A)   Coarse screening – for a spacing of over 3 mm (0.118 in.) or larger (up to 50 mm). For an example in the vegetable canning industry screens sizes do not exceed 5 mm.

B)   Fine screens – Openings of 3 mm or less are classified as fine screens. Fine screens are more applicable to industrial wastewater treatment.

Coarse screens. The most commonly used coarse screens are bar screens and these are typically mounted at an angle of 10–90 degrees to the flow.

Fine screens. The main types of screens are

- Static
- Rotary
- Vibrating.

Static screens are brushed or rundown with vertical bars or perforated plates. Static wedge/curved screens are used to minimise blockage of screens such as in the poultry, meat and other food-processing industries where the particles are sticky and tend to clog the screen. The feed inlet is perpendicular or flows by gravity to the top of the screen and then runs down the side of the screens. The screens have openings of about 1 mm. The liquid drains through the side of the screen. Cleaning nozzles or oscillating sprays are installed to rinse the screens. Removal rates can be in the order of 40–75%.

Figure 8.4 shows a schematic of common curved wire screen.

Rotating screens are more complex but popular within the municipal, poultry and meat industry. They consist of a drum which rotates along its axis, and the effluent enters through an opening at one end. Screened wastewater flows outside the drum. The retained solids are washed out from the screen into a collector in the upper part of the drum by a spray of the wastewater.

The screening media used in these devices is generally of stainless material, with openings varying from 0.7 to 1.5 mm. Materials that dissolve with time such as fish solids need to be screened as soon as possible. By the same token, high-intensity agitation of waste streams (such as pumping or flow-through valves) should be minimized before screening or even settling, since they cause breakdown of solids rendering them more difficult to separate.

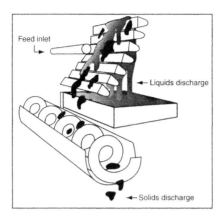

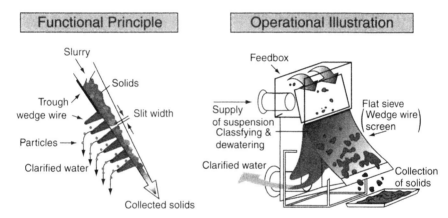

Figure 8.4 A Schematic of a curved screen

Courtesy of Environmental Technology Best Practice Programme, UK

In the Contra-shear design the screens are fed internally to a rotating screen at a tangent. Given the counter-flow direction, higher flow rates than externally fed screens can be achieved. The screens are washed both internally and externally by hot water sprays to remove built up fat and grease. They are suitable for applications with high solids content.

Vibrating screens are used in applications where the moisture content of the solids is low and the wastewater is low in grease content. Typically vibrating screens vibrate at 900–1800 rpm (revolutions per minute).

Weave wire screens such as the on-line self-cleaning filters are able to filter particles up to 5 μm in diameter, that is the maximum particle size required for membrane filtration systems. Refer to Chapter 5 where the application of self-cleaning filters is used in cooling water systems to remove suspended solids.

In the food industry, screens typically achieve a 15–50% reduction in BOD. Capital costs average around A$35 000–150 000.

8.5.2 Sedimentation

Suspended solids may also be removed by settling. Settling is based on the density difference between the suspended particle and the bulk of the liquid which results in the settling of the suspended solids. When the suspended solids are discrete particles (particles do not interact with each other nor bound together) we call this type of settling *sedimentation*.

The terminal or critical velocity (v_c) of discrete particles is a vector between the downward movement of the settling particles and the flow velocity. The particles that reach the bottom of the tank before the outlet will be separated. For discrete settling, calculations can be made on the settling velocity of individual particles. In a settling tank, these move both downwards (settling) and towards the outlet zone with the waterflow as shown in Figure 8.5. v_c can be calculated if the depth of liquid (d) in the tank, the volume of the tank (V) and the flow rate (Q) are known:

$$v_c = d/(V/Q) \tag{8.1}$$

The ratio of V/Q is also known as the residence time of the liquid in the tank. It is called the overflow rate when v_c is expressed in terms of volume of effluent per unit surface area of the tank per unit of time.

Sedimentation is used for the removal of sand, gravel and other dense inert material (specific gravity 2.65) commonly known as grit. Grit is particularly a problem in some wastewaters such as in the fruit and vegetable canning industry.

Surface loading rate is the rate of wastewater flow over the surface area of the sedimentation unit. Typical values are approximately 80–120 $m^3/m^2 d$ (2000–3000 US gpd/ft^2). Primary sedimentation tanks generally provide detention times of 90–150 minutes at average flow rates.

Instead of a settling tank, hydrocyclones can be used for the removal of grit for particles diameter ranging from 5 to 400 μm (0.0002–0.016 in.) [3]. Hydrocyclones utilise specific gravity differentials, tangential flow, angular acceleration to separate particles. The feed enters at the side, at a tangent, the heavier particles spiral down to the bottom of the cone due to centrifugal action. The filtrate and the smaller particles migrate towards the centre, spiral upwards and out through the vortex finder discharging through the overflow pipe. The advantage in hydrocyclones is that they have no moving parts, no extra mechanical energy is required, and unlike screens no clogging

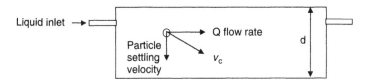

Figure 8.5   A Schematic of discrete settling

takes place. Hydrocyclones can be made from plastic and erosion-resistant materials. However, each cyclone performs only within a narrow band of flow rate. Hydrocyclones are used extensively in the food, mining and pulp and paper industries.

Figure 8.6 shows a graphical illustration of hydrocyclones.

The efficiency with which a hydrocyclone separates at a certain size depends on several design parameters, including [4]:

- diameter of the chamber (D)
- area of the point of entry into the feed chamber ($0.05 \times D^2$)
- area of vortex finder ($0.35 \times D^2$)
- length of the chamber
- angle of the conical section ($10°$ and $20°$)
- on operating parameters such as the flow rate of the input material.

As a general rule, the smaller the diameter of the hydrocyclone, the better it is at separating smaller particles.

Table 8.5 shows the separation of suspended solids in a food-processing plant.

8.5.3 Settling

For solids having a specific gravity less than grit, a longer time is required in settling tanks. Particles that would not settle unaided are referred to as *colloids* and typically measure between 10^{-3} and $1\,\mu m$ (4×10^{-8} and 4×10^{-5} in.) in diameter.

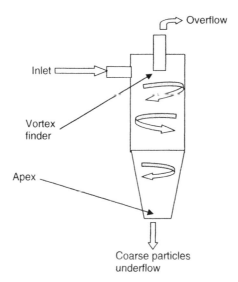

Figure 8.6   A schematic of a hydrocyclone

Table 8.5 Separation efficiency of suspended solids using a hydrocyclone in a food-processing plant

Test	Feed TSS mg/L	Overflow TSS mg/L	Underflow TSS mg/L	Removal efficiency (%)
Run 1	376	37	3,480	90
Run 2	436	51	1,348	88
Run 3	230	73	1,160	68
Run 4	270	12	1,810	96
Run 5	284	16	1,360	94

Courtesy of the Water Management Group Pty Ltd.

For removal of colloidal materials the type of equipment used can be

- conventional horizontal-flow tanks
- tube settlers
- inclined plate settlers
- circular or rectangular clarifiers.

The key design criteria are

- the surface area of the settling tank
- detention time
- tank depth
- surface overflow rate
- weir overflow rate.

Given the inability for laboratory results to exactly mirror actual conditions, settling time is normally doubled and overflow rate is multiplied by 1.5. Typical settling times are around 1–4 hrs. Tank depth is normally around 4 m but can range from 2 to 5 m. After settling the water might of sufficient quality to be reused elsewhere.

8.5.4   Chemically Aided Settling – Coagulation

Otherwise known as coagulation, it is used when colloidal particles do not settle by themselves within a reasonable time and require chemicals to aid in settling. Colloidal particles do not settle because there is an electrical outer layer of negative charge that repel colloids from attaching to one another. To neutralise the outer static charge, chemicals having a high positive charge to mass ratio such as aluminium sulphate, ferric chloride or synthetic coagulants are added. Important polymer properties are polymer type (cationic, anionic and non-ionic), charge density, molecular weight and functional group chemistry. Functional groups can be polyamines or acrylamides. The molecular weight of polymers can range from 10^4 to 10^7.

Charged particles repel each other   Particles bind to each other after charge neutralisation

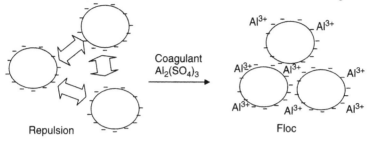

Figure 8.7   How coagulants bind to colloidal particles

Figure 8.7 shows the effect of inorganic coagulants and polymers on colloidal particles. To further aid the process, high molecular polymers or bentonite clay known as flocculants or coagulant aids are added. As Figure 8.7 shows, once the charge is removed the particles are free to collide and stick together forming larger and larger flocs that ultimately become visible to the naked eye and settle quickly. The coagulant aluminium ion precipitates as the $Al(OH)_3$ between pH 6 and 8. The process of chemical coagulation also precipitates metal hydroxides. The resulting jelly-like mass known as *floc* settles to the bottom of the clarifier. The choice of coagulant, coagulant dosage, coagulant aid and pH depends on the wastewater analysis, water temperature, types of clarification equipment and the end use of the treated effluent.

8.5.5 Filtration

Filtration is a separation process that consists in passing a solid–liquid mixture through a porous material (filter) which retains the solids and allows the liquid (filtrate) to pass through. Granular media polishing filters are used for the removal of suspended solids in the 5–50 mg/l range.

The most common example is the conventional sand filter. Table 8.6 shows typical dimensions of filter media. Filters can be classified as

- Single media – sand or anthracite
- Dual media – sand and anthracite
- Multimedia – garnet, sand and anthracite.

Table 8.6 Typical dimensions of multimedia filters

Filter media (Layers arranged from top to bottom)	Specific Gravity	Diameter mm	Height of typical filter layer*
Anthracite	1.6	1–1.5	4.5x
Silica sand	2.65	0.5	2x
Garnet	4.5	0.2–0.4	1x

* Total height of filter media layers – 76 cm (30 in.).

Single-media filters have fine-to-coarse gradation in the direction of flow. After backwashing a single media filter the granular media settles back into place with the coarse materials (largest diameter) on the bottom, fine materials (smallest diameter) on the surface. This natural distribution causes solids to rapidly accumulate at the surface causing increased head-loss and short filter runs. It is preferable to have trapped material accu-mulate more evenly through the depth of the bed to allow longer filter runs.

Multimedia filters employ two or more filter media with different grain size and densities. The media are selected such that the smaller particles are the most dense (e.g. garnet with a specific gravity of 4.5 and particle size of 0.2–0.4 mm), the medium-sized particles have an average density (e.g. sand grains with a specific gravity of 2.65 and a grain size of 0.5 mm) and the largest particles are the least dense (e.g. anthracite grains with a specific gravity of 1.6 and a grain size of 1.0 mm). When mixed and permitted to settle, the multimedia bed will grade itself according to the density of the material. Therefore, the garnet being the densest will settle at the bottom and the anthracite being the least dense will be at the top and sand will be in the middle. When the feed stream flows from top to bottom, the courser-suspended solids will be removed in the upper layers of the filter and the smaller suspended solids near the bottom. Backwashing is performed in the opposite direction. The cost of backwash water and replacement media are the primary expenses of a multimedia filter. Backwash flow rates for multimedia filters are in the order of 24.5–36.5 m³/hr/m² (10–15 US gpm/ft²).

Filter performance is dependent on the amount of turbidity to be removed, the size of the suspended particles and filtration flow rate. With feedwaters of 30 NTU, a multimedia filter is capable of producing effluent with turbidity of 1 NTU and a particulate size in the range of 5–10 μm (0.0002–0.0004 in.). Filtration also reduces insoluble BOD.

Typical filtration flow rates are shown in Table 8.7.

Other filtration methods are cartridge filtration, online filtration (refer to Chapter 5) and membrane filtration. Membrane filtration will be discussed later in this chapter.

Table 8.7 Typical filtration flow rates

Filter rate	Flow rate	
	m³/hr/m²	US gpm/ft²
Single	7.4–9.8	3–4
Dual	14.7	6
Multimedia	14.7–36.5	6–15

8.6 Removal of Fats, Oils and Greases

8.6.1 Sources of Fats, Oil and Grease

Fats, oil and grease (FOG) can be of hydrocarbon, vegetable or mineral based and are found in

- laundry effluent
- food retailing
- processing of vegetables and production of margarine
- metal-finishing and metal-cutting industries
- petroleum refining and petrochemical industries.

Emulsions are defined as the dispersion of liquid in an aqueous medium.

The FOG may be present in these streams as either *free floating* or an emulsion. Emulsions occur when oil droplets are stabilised by contaminants or surfactants giving them a electrical charge that will repel each other, for example mayonnaise. The colour of the emulsion is an indication of the particle size of the oil globules. Table 8.8 shows the relationship between particle size and the colour of the emulsion. Unstable emulsions occur if the droplets coalesce spontaneously at a reasonably rapid rate such as that for oil and vinegar salad dressing. Microemulsions are when the emulsion droplet size is less than $0.05\,\mu$m.

Methods for separating FOG include physical and chemical separation. When both free and emulsified FOG are present, the most economical solution is to first remove as much free oil or fat as possible.

Physical separation can take the form of

- skimming
- plate-type separators
- API (named after the American Petroleum Institute)
- dissolved air flotation system
- membrane systems.

Table 8.8 The relationship between particle size and emulsion colour

Particle size (μm)	Emulsion appearance
Macro globules >150	Two phases may be distinguished – free oil
10–150	Milky white emulsion
10–1.0	Blue-white emulsion
1.0–0.5	Grey semitransparent
< 0.5	Transparent microemulsion, with three or more phases. Stable emulsion. Droplet rise velocity too long for gravity separation.

In chemical separation the dispersed water droplets and solids are destabilised through the use of coagulants to some of these processes to enhance separation and improve quality of the end product.

8.6.2   Free FOG Removal – Skimming

Free oil droplets greater than 150 μm (0.006 in.) separate easily from water by gravity due to differences in specific gravity such as in plate systems, centrifuges as per Stokes Law or differences in surface tension phenomena such as in oil skimmers.

Gravity oil separators are usually rectangular or circular in shape. Basic elements include an inlet distributor, internal baffles and an oil-collecting or skimming device. Size is determined on the basis of rise rate for a critical size oil globule, velocity through the unit and detention time.

The simplest form of gravity separation is the oil and fat interceptor, otherwise known as a *grease trap*. They are commonly found in shopping malls, restaurants and in the food-processing industries where the requirement is only to skim the fat before the effluent is discharged to the sewer. Grease traps are equipped with baffles to enable the grease or fat to float to the surface so that it can be pumped out.

In the petroleum refining industry the API separators are used to capture free oil. The API separators allow the free oil to rise to the surface, allowing it to be collected and sent back to the desalters as *slop oil* for further processing. The API separators are rectangular tanks with a minimum length to width ratio of 5:1 to minimise potential for short circuiting.

A variant of the API separator is the corrugated plate separator (CPI). The CPI is similar to the lamellar clarification for solids removal. Corrugated metal or plastic plates (12–48 plates) are assembled inclined in parallel to each other at distances 2–4 cm (0.75–1.5 in.). The inclined plates allow for coalescence of oil droplets to large oil globules and migrate to the surface and the settled solids migrate downwards. The large oil globules rise to the surface where it is skimmed off. The CPI is considered to be more effective than API units for oil separation.

Hydrocyclones are also effective for oil removal. The principles of operation were discussed under sedimentation.

Skimming of FOG from the surface of separation vessels can be performed on a manual or continuous basis. Suction skimmers and belt skimmers are used for this purpose. Suction skimmers are being replaced by belt skimmers, which are more reliable with low maintenance.

As mentioned earlier, gravity separators are ineffective for oil droplets smaller than 150 μm (0.006 in.).

8.6.3   Emulsified FOG Removal

The removal of emulsified FOG is more difficult. It is the major contributor to high oil concentrations in the effluent streams. For an example, in spent metal-cutting fluid the free oil can be 5–10% whereas the emulsified oil fraction can be 80–90% of the total. Emulsified oil separation technologies include air flotation, membrane and adsorption processes.

8.6.3.1   Air Flotation

Flotation is basically sedimentation in reverse, to remove floatable materials and solids by increasing the density difference between the oil emulsion and the water phases. In the air flotation process, the feed water is saturated with air under pressure and then expanded through a restriction to atmospheric pressure. The sudden expansion forms microscopic air bubbles, which attach to oil droplets. The air/oil droplets have a lower density than oil micelles and micro-emulsions causing them to rise to the surface much more quickly where they are continuously skimmed off and removed. Air flotation is particularly useful when treating waters which are high in total suspended solids (TSS) or have highly variable suspended solids content.

There are two types of air flotation systems in use:

1.   induced air flotation (IAF)
2.   dissolved air flotation (DAF).

In IAF systems, air is introduced into the wastewater by an aspirator device or by a fine hole diffuser to form bubbles up to $1000\,\mu m$ (0.04 in.) in diameter. Since the air/water contact occurs essentially at atmospheric pressure, air bubbles are entrained in the water rather than being formed in the water. Even though the removal efficiencies are less effective than for DAF it is cheaper since no air pressurisation is required and can remove larger amounts of oil/solid flocs which are brought to the surface.

The DAF process is more common than the IAF process. Figure 8.8 shows a schematic of a DAF process. In the DAF process, due to the air bubbles being formed inside the pressurised chamber and after the pressure control valve (expansion valve), microscopic air bubbles and air/oil bubbles $30–120\,\mu m$ (0.0012–0.048 in.) in size are formed. The air/water stream enters the bottom of the tank. Initial separation occurs at this point. Typical air pressures are 300–600 kPa (44–85 psig). Chemical coagulants such as ferric chloride ($FeCl_3$) or synthetic proprietary products known as reverse emulsion breakers are added to enhance the separation process. However, in food-processing plants, the addition of synthetic coagulants renders the sludge unsuitable as animal stock feed.

There are three methods of aeration:

1.   aeration of the main stream
2.   aeration of the side stream
3.   aeration and recycling a slip stream of clean effluent.

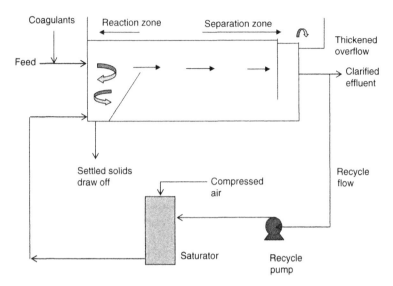

Figure 8.8   A schematic of a dissolved air flotation process

Aeration of the main stream is not common practice. It is generally used when chemical treatment is not required since the floc formed will be destroyed when it goes through the expansion chamber.

In the aeration of a recycled stream, 15–50% of clean effluent is more advantageous than aeration of slip stream, since it does not contribute to dilution of the feed water (if potable water used as the slip stream) and allows for the total influent flow to be flocculated.

There are essentially three types of flotation designs:

1.   rectangular
2.   circular
3.   corrugated plate pack.

Circular flotation tanks and corrugated plate designs are preferred when treating water with high suspended solids concentrations for a given unit capacity.

The DAF units can reduce oil concentrations to 10–25 mg/L in the effluent as long as the influent oil concentrations are less than 160 mg/L [5, 6]. The DAF systems operate at higher hydraulic loading rates (5–15 m/h [2–6 gpm/ft^2]) than gravity sedimentation equipment and consequently detention times are shorter (about 15–30 minutes). Therefore DAF systems require less space.

The key design and operating figures are given in Table 8.9. Performance of DAF systems are dependent on several factors including:

*   the solids concentration – higher solids content usually gives higher removal efficiencies

Table 8.9 Key design and operating parameters of DAF systems

Parameter	Metric	US units
Detention time	15–30 minutes	15–30 minutes
Solids loading rate (determines the size of the unit)	2–25 kg/m^2 h	0.5–5.0 lb/h ft^2
Overflow rate (surface hydsraulic loading rate)	5–12*	2–5*
Recycle Ratio	15–50%	15–50%
Saturator pressure	450–600 kPa	44–85 psig
Air to solids ratio	0.005–0.06 mL/mg	0.005–0.06 mL/mg

* A key consideration with regard to this design parameter is whether the loading rate includes the recycled volume as well as the influent wastewater volume being applied per unit area of the system.

- air to solids (A/S) defined as the amount of air released after pressure reduction and the amount of solids present in the wastewater. There is usually an optimum A/S which is determined by bench scale tests.

Key factors in the successful operation of DAF units are the maintenance of proper pH (usually between 4.5 and 6, with 5 being most common to minimise protein solubility and break-up emulsions), proper flow rates and the continuous presence of trained operators.

8.6.3.2 Ultrafiltration

Membrane processes, especially ultrafiltration (UF) membranes, can be used for the removal of emulsified oil. Ultrafiltration is a low pressure membrane process (100–1000 kPa [14.5–145 psig]) that separates suspended solids and high molecular weight dissolved solids from liquid. The separation is performed by a semipermeable membrane. Low molecular weight dissolved solids such as salts and surfactants pass through the membrane and are discharged with the permeate while high molecular weight dissolved solids >3000 molecular weight (oil, colloidal solids and suspended solids) are concentrated and rejected. Recovery rates of 75% can be achieved with oil concentrations reduced from 3500–30 mg/L, a 99% reduction in the concentration of oil.

Aside from UF, other membranes such as microfiltration and nanofiltration can also be used to separate emulsions. Table 8.10 shows a specific type of plate membrane process known as VSEP (vibratory shear enhanced membrane process) [7] which has achieved high removal efficiencies.

Other methods of treatment are

- biological treatment
- centrifugation
- evaporation
- activated carbon

Biological treatment will be discussed in the next section.

Table 8.10 Oil removal efficiencies using the VSEP process

Process	Membrane	% Recovery	Initial concentration solids %	Final % solids in reject stream
Lubricant wastewater	100,000 MWCO* UF	60	10.33	25.82
Lubricant wastewater	Nanofiltration	75	2.37	37.02
Machine coolant	7,000 MWCO UF	75	2.89	13.82
Oily wastewater	Nanofiltration	80	0.07	0.81
Oily wastewater	70,000 MWCO UF	60	0.15	1.47
Oily wastewater	Nanofiltration	90	0.61	6.64
Produced water/Silt	100 k MWCO UF	70	22.69	84.19
Washwater degreaser	Reverse Osmosis	60	3.02	9.59

Adapted from New Logic Research Inc. *Using Vibrating Membranes to treat oily wastewater.*
* MWCO refers to molecular weight cut off which is a unit to measure the porosity of the UF membrane. The higher the MWCO, the greater the diameter of the pore. These aspects will be discussed under membranes.

Centrifugation and evaporation are energy-intensive processes. Evaporation is suitable for small volumes of poorly emulsified oils.

Evaporators work on the principle that by raising the temperature to of the wastewater 70° C–80° C (160–176° F), the emulsion will be broken. Water will evaporate while oil is left behind. This method is not effective if solids are too high or if the emulsion cannot be broken by heat.

Activated carbon is used to remove traces of oil (1 mg/L or less), to achieve ultrapure water or to remove the surfactants in the wastewater. Activated carbon will be discussed under adsorption.

8.7   Removal of Biodegradable Organics

For the vast majority of industries, removal of biodegradable organics (BOD) is a prerequisite before the effluent streams can be reused. Their removal is also important if discharging to the public sewer, and in the food industries high BOD can indicate product losses.

Many water utilities accept trade waste discharges subject to concentrations and mass loadings being lower than the acceptance standard or as specified in the trade waste agreement. The BOD concentrations higher than the acceptance standard are charged at higher rates since high BOD concentrations in the sewer contribute to rapid corrosion of sewer.

Figure 8.9 shows Sydney Water's trade waste acceptance standards, domestic equivalents and charging rates for domestic substances. In the UK the Mogden formula is used to calculate trade waste strengths.

Substance	Acceptance standard (mg/L)	Domestic equivalent (mg/L)	Note	$ / kg
Suspended solids	600	200		0.765
BODs - primary treatment		230	1	0.1069 + {0.0173 x (BOD/600)}
BODs - secondary treatment		230	1	0.603 + {0.0173 x (BOD/600)}
Grease - primary treatment	110	50	2	1.079
Grease - secondary treatment	200	50	2	1.079
Ammonia	100	35	3, 5	1.789
Nitrogen	150	50	4	0.151
Phosphorus	50	10	4	1.196
Sulphate	2000	50		0.118 x (SO₄/2000)
Total dissolved solids (ocean systems, no discharge limitation)	10000	450		0.005
Total dissolved solids (inland & ocean systems with limitations)	500	450		0.005
Total dissolved solids (inland & ocean systems with advanced treatment to remove TDS)	10000	450		0.061

Figure 8.9   Trade waste acceptance standards, domestic equivalents and charging rates for domestic substances

Courtesy of Sydney Water – Trade Waste Industrial Customers – Acceptance Standards and Charging Rates 2006–2007.

In many food-processing plants, high BOD discharges represent product loss to the sewer. For instance, in the dairy industry 1 kg of BOD represents 9 kg of milk. Therefore, minimising product losses will reduce trade waste charges while increasing profits. The following example illustrates the point.

Worked example

A dairy producing 500 000 L milk/day has a sewer usage discharge factor of 0.88 and discharges 440 kL/day. The BOD of the trade waste is 2500 mg/L. The BOD of raw milk is 104 600 mg/L. What are the trade waste charges and value of lost milk? If the BOD was reduced to 1000 mg/L what are the new charges? Assume that raw milk costs $0.10/L and the trade waste is discharged to a secondary treatment plant. Use the values in Figure 8.9 for acceptance standards and charging rates.

	Before waste minimisation	After waste minimisation
Loss of raw material, kg/annum	$2,500 \times 440 \times 1,000 / 104,600 = 10,516$ kg/day $\times 365$ days $= 3,838,432$ kg. $= 2.1\%$ loss	$1,000 \times 440 \times 1,000 / 104,600 = 4206.5$ kg/day $\times 365$ days $= 1,535,373$ kg $= 0.8\%$ loss

Cost of lost raw, material/annum	$ 383,843	$153,573
BOD, $/kg	$0.603 + (0.0173 × 2500/600) = $0.675/kg	$0.603 + (0.0173 × 1000/600) = $0.632/kg
Annual trade waste charges	= $1100/day = $401,500	= $440/day = $160,600

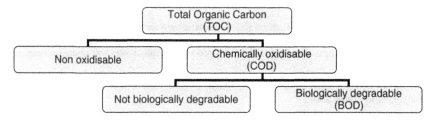

Figure 8.10   The relationship between the organic carbon fractions in wastewater

Organic matter can be classified as shown Figure 8.10.

As Table 8.3 illustrated, when the BOD/COD ratio is greater than 0.4 then the effluent is generally biodegradable, provided that there are no toxins present.

Microbes degrade organics in nature and these principles form the basis of biological treatment albeit, at very high concentrations. Some inorganic compounds such as ammonia, cyanide, sulphide, sulphate and thiocyanate are also biodegradable.

Biological treatment methods can be classified as

- aerobic (using oxygen as the electron acceptor)
- anaerobic (using other electron acceptors such as sulphate, phosphate or other organics other than oxygen).
- hybrid (*combination of aerobic and anaerobic processes*).

The aerobic and anaerobic processes as shown graphically in Figure 8.11 produce different end products. Aerobic processes produces sludge and heat while the anaerobic process produces sludge and methane gas, which can be utilised to generate renewable energy.

The various types of biological processes are shown in the table below. Many of these processes are discussed in wastewater treatment literature and therefore, in the interest of space, will be discussed only briefly in this chapter.

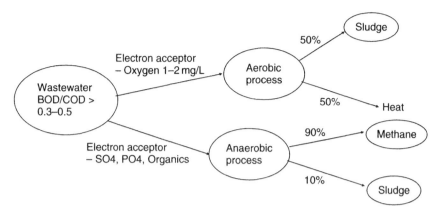

Figure 8.11   An illustration of aerobic and anaerobic processes

Aerobic	Anaerobic	Hybrid
Biological filtration	Anaerobic digestion	Facultative lagoon
Activated sludge (ASP)	Hybrid anaerobic	
Oxidation ditch	Anaerobic lagoon	
Sequencing batch reactor (SBR)	UASB	
Spray irrigation	Anaerobic filters	
Rotating biological discs	Continuous stirred tank reactor (CSTR)	
Submerged Membrane bioreactor		

Facultative lagoons are the most common form of aquatic treatment-lagoon technology currently in use. The water layer near the surface is aerobic while the bottom layer, which includes sludge deposits, is anaerobic. They are applicable when there is sufficient land area.

Aerobic processes can be divided into two groups. These are

1. *suspended growth system*
2. *attached growth system.*

In the suspended growth systems the biomass is maintained in suspension in an aqueous environment. Oxygen transfer occurs directly as dissolved oxygen – from the aqueous phase to the biomass. The most representative and flexible of these processes is the **activated sludge process** (ASP). It must be noted that not all biological matter is digested by the bacteria and these are called 'hard' BOD. It may take days for hard BOD to be digested.

In the attached growth systems the biomass is supported by a solid phase on which it grows. Oxygen is transferred directly to the biomass from the gaseous phase, that is air. The **trickling filter** is the best example of this class.

Trickling filters consist of a fixed film biological reactor, which is followed by a secondary settling tank.

The ASP will be examined, given its ability to handle a variety of industrial wastewater streams. For this reason, it is quite common in industry.

8.7.1 Activated Sludge Process

The ASP was developed in the United Kingdom in the early years of the twentieth century and now forms the centrepiece of the biological treatment processes worldwide.

The ASP as shown in Figure 8.12 consists of a balancing tank to equalise flow and homogenise effluent loads, a fluid bed reactor known as the *aeration tank* where the organic load is stabilised, a secondary settling tank called the *clarifier* where the biological mass is allowed to settle to the bottom of the clarifier while the clear liquid above it is filtered and discharged to the sewer or to the receiving waters. To maintain the bacterial population in the aeration tank in sufficient concentration, a portion of the settled activated sludge (about 25–50%) is pumped or recycled back to the aeration tank. The rest of the sludge is sent to a thickener for dewatering.

The heart of the process is the aeration basin, which is a complex ecosystem of competing organisms. The dominant organisms are the 300 species of unicellular bacteria [8]. Bacteria can be classified as per Table 8.11 by shape and function.

The other organisms present in settleable solids are fungi, protozoa (also single cell organisms), rotifers and sometimes nematodes (worms). Common types of protozoa found in activated sludge mass are the free swimming variant known as 'ciliates', amoeba and flagellates. Rotifers are multi-cellular animals having a digestive system and rotating cilia around their head and hence the name. Nematodes can be 0.2–1 mm in length.

Their presence or absence is indicative of the state of the biomass and age. For example, the presence of filamentous bacteria indicates bulking sludge. Similarly, the presence of free swimming protozoa (ciliates) is a sign of a healthy biomass with adequate oxygen supply and the presence of nematodes and rotifers in large numbers is a sign of long and mature sludge.

The process requirements for ASPs are similar to other natural biological processes. The ASPs are capable of reducing organic load by 85–95% with an aeration time of about 6 hrs. However, successful operation of these systems requires trained manpower and the conditions conducive to biological growth. These are

- nutrient loading
- dissolved oxygen (DO)

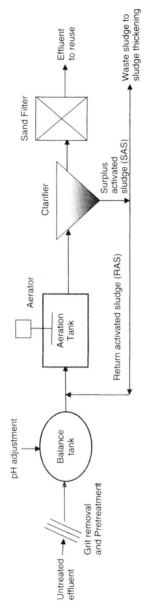

Figure 8.12   A schematic of a typical activated sludge process

Table 8.11 Classification of Bacteria

By shape	By function	Comments
Spherical	Floc formers	Majority. Floc formers are desirable and required for biodegradation to establish a stable flow with good settling characteristics. Size is about 0.5–2 μm (0.00002–0.00008 in.).
Rod	Filamentous	Undesirable. Cause of bulking and foaming in aeration tanks. They comprise of long chains of small bacterial cells and can reach lengths of 100 μm(0.004 in.).
Spiral	Nitrifiers	Desirable. Bacteria to convert ammoniacal nitrogen to nitrate through a two-step process. First step is the conversion of ammonia to nitrite and then the nitrite is converted to nitrate.

- pH
- temperature
- the presence of toxic or inhibitory materials.

Nutrient loading – ASPs operate best when the organic load and flow is relatively constant. Unfavourable conditions can be tolerated as long as there are no sudden fluctuations in flow and organic loading which lead to unsteady state biomass conditions in ASPs. For example, after a rain the stormwater oil concentrations in oil refineries can exceed several thousand mg/L. This quantity of oil entering the ASP can kill the bacteria. Therefore, a balancing tank is a prerequisite in an ASP to even out the flow and organic loading.

The BOD/COD ratio gives an indication of its biodegradability and nutrient requirements. F:M ratio is one of the most important parameters to monitor the relationship between the influent nutrient quantity (equal to the pollutant quantity) and the existing mass of micro-organisms in the aeration tank available to treat the incoming BOD.

For conventional plants the F:M is around 0.2–0.5. At higher values while the rate of treatment increases, the sludge has poor settling characteristics. F:M values below 0.2 are associated with slow BOD removal rates, but with very good sludge settlement.

High load	0.8–1.5	F:M per day
Normal load	0.3–0.7	F:M per day
Low load	0.05–0.2	F:M per day

Once the BOD/COD ratio for the wastewater has been established, then COD values can be used to calculate F:M ratios.

Sludge age is an indication of the F:M ratio. Shorter times are indicative of high growth rate and conversely a longer time is indicative of low growth rates. For instance, at high BOD loading the sludge age will be less than

2 days. For plants requiring nitrogen removal sludge age is normally around 10–15 days [9].

Besides carbon, hydrogen and oxygen the biomass requires nitrogen (N), phosphorous (P) and micronutrients such as iron, calcium, magnesium, copper, zinc and so on. Unlike domestic sewage which has a C:N:P = 100:17:5–100:19:6, most industrial wastewaters such as paper mill effluent, brewery effluent lack N and P which should be added (in the form of urea, superphosphate or ammonium phosphate) to maintain optimal conditions. The minimum BOD to N and P ratio required for optimal microbial growth in the aeration tank is given as:

$$BOD:N:P \quad = \quad 100:5:1 \tag{8.2}$$

Brewing and some food-processing industries are particularly conducive to sludge bulking probably due to nutrient imbalance.

Dissolved oxygen (DO) – As mentioned earlier aerobic bacteria require DO to produce energy. The minimum DO concentration required for ASP is 0.5 mg/L. Therefore, oxygen is controlled at 1–2 mg/L.

The DO consumption depends on various factors, including the F:M (sludge loading). For a given BOD, more DO is required as the sludge loading decreases. At a low sludge loading, large quantities of DO are required in order to break down the activated sludge.

Nitrification also occurs at low sludge loading, considerably increasing the DO demand. For satisfactory nitrification, the DO needs to be maintained at around 4 mg/L.

The DO is supplied by mechanical or diffused aeration systems such as surface aerators, motor-driven turbine, spargers, swing diffusers, through dispersed aeration and pure oxygen injection. Typically 0.5–2.0 kg of DO is produced per kWh. Aeration is the most expensive operating cost in an ASP and needs to be optimised by monitoring the oxygen concentration.

pH – The optimum pH for ASP is within 6.5–9.5 and for nitrification 7.0–8.5.

Temperature – The typical temperatures for operation of ASPs are 5° C–30° C. But some thermophilic species of bacteria can operate at higher temperatures as high as 60° C. For every 10° C rise in temperature the growth rate doubles (Arrhenius rule). Higher the temperature, the critical oxygen concentration also increases.

Common problems in ASPs are shown in Table 8.12.

There are other variations of this, such as step variation, contact stabilisation, high rate, extended aeration and pure oxygen systems. These are well documented in literature.

For wastewater streams with *non-steady-state* conditions or when the available foot print is limited, the Sequencing Batch Reactor (SBR) process is a viable option. It is an activated sludge process designed to operate under

Table 8.12 Common problems in ASPs.

Problem	Possible cause
High solids content in clarified effluent	1. Too high or too low solids retention time 2. Presence of large number of filamentous bacteria. Treat with biocides.
Settled sludge rises back again	Denitrification produces nitrogen gas that becomes trapped in the sludge causing it to float.
Bulking sludge does not settle quickly	Presence of filamentous micro-organisms. Treat with biocides.
Poor organic load reduction	1. High sludge wastage 2. Lack of N and P 3. Short circuiting in the clarifier 4. Low DO 5. Presence of toxic substances.
Odour	Anaerobic conditions present in the clarifier or low DO.
Foaming (mouse like)	Presence of a particular type of filamentous bacteria called *Nocardia*. Treat with biocides.

non-steady-state conditions. An SBR operates in a true batch mode with equalisation, aeration and sludge settlement occurring in the same tank. To optimise the operation, two or more batch reactors are used in a predetermined sequence. The SBR system can be designed with the ability to treat a wide range of low to intermittent influent volumes, whereas the continuous system is based upon a fixed influent flow rate. Once the reactor is full, the flow is discontinued and the aeration and mixing is discontinued after the biological reactions are complete. The biomass settles and the treated supernatant is removed. Excess biomass is wasted at any time during the cycle. Thus, there is a degree of flexibility associated with working in a time rather than in a space sequence. The SBRs produce a BOD removal efficiency of 85% depending on the mode of operation. Additional advantage of SBRs are that the cost of clarification equipment is avoided. The negatives of SBRs are that the control mechanisms are more complicated than conventional activated sludge systems.

Typical SBR effluent water quality parameters are

TSS $<10\,mg/L$
BOD $<10\,mg/L$
Total N $5–8\,mg/L$
Total P $1–2\,mg/L$

8.7.2 Anaerobic Processes

Anaerobic processes are best suited to treat high concentrations of BOD such as wastewater from olive oil processing which can contain BOD concentrations as high as 30 600 mg/L and COD concentrations of 97 000 mg/L, beer and wine distillery effluent, textile mill effluent–containing azo dyes, fish-processing effluent and other hard-to-process effluent. Aerobic systems are unsuitable under these high BOD conditions.

Anaerobic processes contain large numbers of highly specialised bacteria which, in the absence of oxygen, convert the concentrated organics into methane gas (CH_4) and carbon dioxide (CO_2) rather than to new cell growth.

One of the most popular anaerobic processes is *The Upflow Anaerobic Sludge Blanket* (UASB). In the UASB reactor, the wastewater enters a vertical tank at the bottom. The wastewater passes upwards through an anaerobic sludge bed where the micro-organisms in the sludge come into contact with wastewater-substrates. The sludge bed which occupies 30–60% of the reactor volume is composed of micro-organisms that naturally form granules (pellets) of 0.5–2 mm diameter that have a high sedimentation velocity and thus resist wash-out from the system – even at high hydraulic loads.

Figure 8.13 shows a graphical illustration of the UASB process.

The gas produced from a well-functioning anaerobic plant contains 60–70% methane. The rest is CO_2 and insignificant amounts of nitrogen and hydrogen gas. The upward motion of released gas bubbles causes hydraulic turbulence that provides reactor mixing without any mechanical parts. At the top of the reactor, the water phase is separated from sludge solids and gas in a three-phase separator – also known the *gas–liquid–solids separator.*

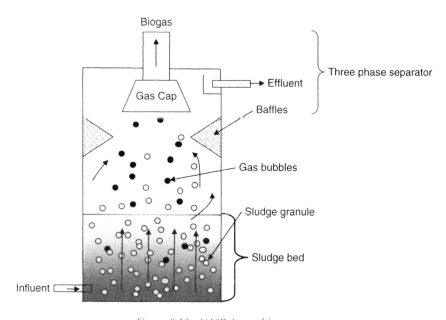

Figure 8.13   UASB Anaerobic process

8.7.2.1 Performance

Operational performance indicators for the UASB process is as follows:

COD removal efficiencies	80–95%
Hydraulic loading rate (HRT)	0.25–1.25 days
Solid retention time	50–100 days or more.
Energy production	1.27×10^7 J/kg (5470 Btu/lb COD) [10]

Operational costs are reported as low as US$0.10/kg COD (due to the generation of own power).

The advantages of anaerobic processes over aerobic processes are summarised in Table 8.13.

Table 8.13 **Comparison of Aerobic (ASP) and Anaerobic Treatment (UASB) processes**

Criteria	ASP	UASB
Applicability	Limited to BOD loadings less than 2000 mg/L. High strength wastes > 2000 mg/L requires dilution [10].	Ideal for high organic strength wastewater. 5–20 kg COD/m^3.d as against 0.5–3 kg/m^3.d for aerobic systems [11]
	Suitable for cold weather operation. Not suitable for high temperatures.	More suited for temperatures above 22° C. Can operate at temperatures as high as 57° C.
	Tolerates a reasonable amount of toxic substances	Cannot tolerate toxic substances
	High nutrient requirements C : N : P = 100 : 5 : 1	Low nutrient requirement C : N : P = 250–350 : 5 : 1
	Continuous operation. Does not tolerate long shutdowns	Applicable to plants with seasonal production.
	High space requirements. Requires a number of tanks.	Low space requirements due to high strength of organic waste.
	Small plants are economically feasible	Small plants not economically viable.
Treatment efficacy	Relatively short sludge retention time is required	Long sludge retention time is required
	N and P removal achieved	Little Nor P is removed.
	Start up time is short	A slow process. Can take months for start up.
	High sludge volumes are produced	Low sludge volumes are produced.
	No flaring of gas is required	Flaring or combustion of methane gas is required.
Costs	Medium to high capital cost.	Medium to high capital cost
	High operational costs. High energy costs for aeration, maintenance of aerators, transport costs for sludge disposal and nutrients costs.	Low operational costs. Low power costs due to generation of methane gas. Low nutrient costs. Low sludge disposal transport costs.

8.7.3 Membrane Bioreactors

Membrane bioreactors (MBR) were originally conceived in the 1960s. MBRs are now becoming popular in domestic and industrial wastewater treatment plants with BOD levels as high as 18 000 mg/L. MBRs are ideal for users who have space constraints, require a consistent quality of low BOD and TSS effluent.

In an aerobic ASP plant, the rate limiting stage is the settling process in secondary clarifier. It limits the solids in the aeration tank to 1500–5000 mg/L (mixed liquor suspended solids (MLSS)). Since the MBR process does not rely on gravitational settling for separation of solids from the liquid, it can operate at significantly higher MLSS concentrations. It relies on filtration performance of microfiltration and ultrafiltration membranes to separate the solids greater than 0.4 μm (0.00016 in.) which includes many types of bacteria and viruses. Therefore, the effluent quality is independent of the settling characteristics of the sludge but MBRs are limited by the fluid dynamics of high-strength suspended solids and oxygen transfer. Typical MLSS strengths are in the order of 10 000–12 000 mg/L. The secret to the capability of the membranes to perform in a biological reactor in a high solids environment is due to the fact that the membranes are used to contain the biomass in the reactor while discharging a purified stream.

The MBR systems are of two designs – submerged and external recirculation. Figure 8.14 shows a schematic of a submerged MBR process.

In the submerged MBR, the microfiltration or ultrafiltration membranes are submerged in the aeration tank. As illustrated, pumps draw product water and sludge. A portion of the sludge is recycled. Air blowers supply air to scour the membranes and for uniform distribution of suspended solids. In the external MBR, the membranes are external to the aeration tank.

The foot print of the bioreactor in an MBR is one-third to one-fifth of that of a conventional ASP. Hydraulic retention times are typically 4–20 hrs with sludge retention times in the order of 15–45 days and recovery is 99%. The long sludge age provides adequate time to reduce excess biomass

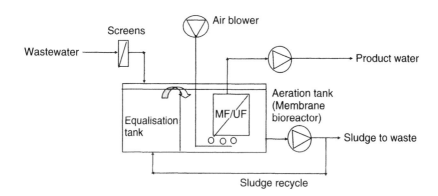

Figure 8.14 A schematic of a submerged MBR

production. The disadvantages of MBRs are that they are prone to intensive fouling and therefore need to be cleaned frequently. Cleaning is normally accomplished by air scouring every 13–15 minutes and cleaning with high concentrations of sodium hypochlorite solutions is carried out once every 2–3 months.

The immersed membranes are less energy intensive compared to external MBRs. However, they use more membrane area and operate at lower flux levels. These terms are explained later in this chapter.

The effluent can be used as reverse osmosis feed water without further treatment.

Case Study: MBR Application in a Dairy

A dairy installed an MBR plant to treat their wastewater effluent from the processing of milk powder, whey, cheese and bacon. The plant has a flow rate of $2000\,m^3/d$ (367 US gpm)

Process Data

		Inlet	Outlet
COD	mg/L	3,600	<50
BOD	mg/L	2,250	<3
Ammonia	mg/L	75	<1
TKN	mg/L		<15
Total P	mg/L		<0.5

8.8   Removal of Heavy Metals

Heavy metals are present in natural waters as well as in industrial wastewaters. Metals such as cadmium, arsenic (not a metal but categorised as one), nickel, copper, lead, chromium discharges to receiving waters are heavily regulated to ensure that concentrations in the receiving waters do not exceed safe limits set by regulatory authorities – such as the US EPA regulations (Federal Hazardous Waste Regulations 40 CFR 261). While chromium is the most widely used heavy metal that is discharged to the environment, *cadmium, lead and mercury* are the most toxic metals to living organisms. Arsenic is found in many well waters in countries such as in Bangladesh. For these reasons, water authorities also impose strict acceptance standards at the entry point to the sewer for the same reason. The sample of metals, the industries where they are typically found, and their acceptance standards to the public sewer are shown in Table 8.14.

Table 8.14 A sample of industries discharging heavy metals

Metal	Industry	Acceptance standards* mg/L
Cadmium	Plating, glassware	1
Chromium	Plating, chrome tanning, alum anodising, ceramics, porcelain, textile dyeing	3
Cobalt	Metal plating, foundries	5
Copper	Foundries, copper plating, automotive industry	5
Iron	Foundries, metal plating, dye manufacture	50
Lead	Ceramics, metallurgical industry, nonferrous smelting, mining, chemicals industry	2
Mercury	Metallurgical industry, alloying, wood preservative, chlorine manufacturing	0.03
Nickel	Plating, oil refining	3
Selenium	Paint manufacture, carpet manufacture, electronics production	5
Uranium	Nuclear industry	10
Zinc	Galvanising zinc plating, metal mining, foundries, porcelain, paint manufacture	5

* Adapted from Sydney Water's Trade Waste Policy. www.sydneywater.com.au.
Courtesy of Sydney Water.

8.8.1 Chemical Precipitation

The principal method used to remove heavy metals is by precipitation of the metal ion as the metal hydroxide by the addition of lime or caustic to adjust the pH to reduce the solubility of the dissolved metals. Figure 8.15 shows the metal hydroxide solubility curve showing the solubility of the common heavy metal ions and their solubility versus pH.

Removal of metals by precipitation consists of the following steps:

1. Pretreatment
2. Precipitation
3. Flocculation
4. Settling.

1. Pretreatment step
To aid the precipitation process and achieve even lower concentrations than the predicted theoretical values, alum (aluminium sulphate) or ferric chloride is added to co-precipitate with the heavy metals as aluminium hydroxide $[Al(OH)_3]$ or ferric hydroxide $[Fe(OH)_3]$.

Chromium is present in both the trivalent (Cr^{3+}) and the hexavalent Cr^{6+} state. In process solutions and wastes, the dominant species is Cr^{6+}. Reducing agents are added to convert hexavalent Cr^{6+} to Cr^{3+}. Unlike most heavy metals which are precipitated readily as insoluble hydroxides by pH adjustment, Cr^{6+} must first be reduced to Cr^{3+} because it forms the chromate complex which behaves as an anion and cannot form an insoluble hydroxide. The

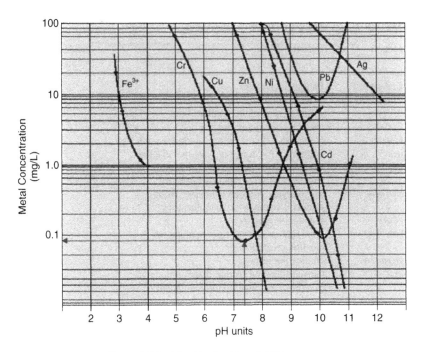

Figure 8.15   Metal hydroxide solubility curve

most commonly used reducing agents are sulphur dioxide gas and sodium metabisulphite (dry granular power).

Amphoteric metals such as copper, cadmium and zinc readily form ammoniacal complexes and under these circumstances the metal hydroxide method is ineffective. Ammonia complexes may be destroyed by oxidation, co-precipitation with ferrous or ferric chloride or one of the most cost-effective methods is to break the ammoniacal complex by the addition of sodium sulphide.

2. Precipitation step

This is carried out by the addition of alkalis sodium hydroxide (NaOH), calcium hydroxide or hydrated lime, $[Ca(OH)_2]$. Other alkalis used are magnesium hydroxide $[Mg(OH)_2]$, calcium chloride $[Ca(Cl_2)]$, sodium carbonate (Na_2CO_3) and sodium bicarbonate $(NaHCO_3)$. Hydrated lime is cheaper than NaOH and produces a metal hydroxide precipitate with faster settling rates and more amenable to dewatering. On the other hand, NaOH reacts faster, has a simpler dosing system and produces less sludge. The disadvantage of the hydroxide method is that ammoniacal complexes and highly chelated metal complexes are not precipitated.

Sulphide precipitation is an alternative method to hydroxide precipitation which precipitates metals as sulphides instead of hydroxides. As Table 8.15 shows metal sulphides have a significantly lower solubility than the corresponding hydroxide.

Table 8.15 Metal hydroxide and sulphide solubilities in mg/L

Metal	Valency	As Hydroxide	As Sulphide
Cd	2+	2.3×10^{-5}	6.7×10^{-10}
Cr	3+	8.4×10^{-4}	No precipitate
Co	2+	2.2×10^{-1}	1.0×10^{-8}
Cu	2+	2.2×10^{-2}	5.8×10^{-18}
Iron	2+	8.9×10^{-1}	3.4×10^{-5}
Lead	2+	2.1	3.8×10^{-9}
Mn	2+	1.2	2.1×10^{-3}
Mercury	2+	3.9×10^{-4}	9×10^{-20}
Ni	2+	6.9×10^{-3}	6.9×10^{-8}
Ag		13.3	7.4×10^{-12}
Tin	2+	1.1×10^{-4}	3.8×10^{-4}
Zinc	2+	1.1	2.3×10^{-7}

In well-designed chlor-alkali plants this method reportedly achieves mercury removal efficiencies of 95–99% [12]. Therefore sulphide precipitation is one way to meet strict limits on dissolved metal concentrations even in the presence of chelating agents. Sulphide precipitates also tend to be less hydrated and therefore less voluminous than hydroxides. Even in the presence of 100 mg/L of EDTA, a chelating agent, copper can be reduced to 1 mg/L with sulphide precipitation. The drawbacks of this method are that hydrogen sulphide can be generated at low pH, excess sulphide is a pollutant and that metal sulphides produce a floc which is more difficult to dewater. Also the metals may leach out from the sludge. For these reasons, sulphide precipitation is used more as a polishing step after hydroxide precipitation [3] Two processes are used for sulphide precipitation: the soluble sulphide process uses sodium sulphide as the treatment reagent, and the insoluble sulphide process uses ferrous sulphide. The sparingly soluble ferrous sulphide overcomes the problem of excess sulphide dosage.

Common target for metal precipitation is a pH between 8.5 and 10. In practice the pH rarely exceeds 9.2. As Figure 8.16 shows the metal hydroxide solubilities, the lowest point of each curve corresponds to the pH at which that metal species will be removed to its minimum solubility point. For an example, as Figure 8.16 shows, the lowest solubility for chromium (Cr^{3+}) is at a pH of 7.5 where the theoretical solubility is 0.08 mg/L. Chromium being an amphoteric metal at pH values below or higher than 7.5 is more soluble.

When two or more metals are present and have different pH at which the minimum solubility occurs, an optimum pH is selected. For example, if both chromium and nickel are present then the optimum pH value to precipitate both metals are at a pH of 9.0–9.5 even though nickel has a minimum solubility at pH 10.2 and chromium at 7.5. Another solution is to use a class of compounds known as dithiocarbamates or trithiocarbamates. These provide precipitation even in the presence of chelating agents such as EDTA.

Barium is an exception where it is commonly precipitated as the sulphate ($BaSO_4$).

3. Flocculation

For faster precipitation polyelectrolytes consisting of organic chemical flocculants are added. These are similar to the ones discussed previously under chemical settling.

4. Clarification

Clarifiers and plate separators are used to settle the sludge followed by media filtration as a polishing step.

Disadvantages of metals precipitation may include the following.

- The presence of multiple metal species may lead to removal difficulties as a result of amphoteric natures of different compounds (i.e. optimisation on one metal species may prevent removal of another).
- As discharge standards become more stringent, further treatment may be required.
- Metal hydroxide sludges must pass EPA requirements prior to land disposal.
- Soluble hexavalent chrome requires extra treatment prior to coagulation and flocculation.
- Reagent addition must be carefully controlled to preclude unacceptable concentrations in treatment effluent.
- Efficacy of the system relies on adequate solids separation techniques (e.g. clarification, flocculation and/or filtration).
- Process may generate toxic sludge requiring proper disposal.
- Process can be costly, depending on reagents used, required system controls, and required operator involvement in system operation.
- Dissolved salts are added to the treated water as a result of pH adjustment.
- Polymer may need to be added to the water to achieve adequate settling of solids.
- Treated water will often require pH adjustment.
- Metals held in solution by complexing agents (e.g. cyanide or EDTA) are difficult to precipitate requiring more expensive solutions.

8.8.2   Ion Exchange

Another method used for removal of heavy metals is ion exchange. Ion exchange is used widely in the metal finishing industry to remove dissolved heavy metals from the rinse water such as for the removal of nickel, copper, tin, aluminium anodising and zinc ions.

Ion exchange principles are discussed in Chapter 7.

In most plating processes, water is used to cleanse the surface of the parts after each process bath. To maintain quality standards, the level of dissolved solids in the rinse water must be controlled. Fresh water added to the rinse tank accomplishes this purpose, and the overflow water is treated to remove pollutants and then discharged. As the metal salts, acids and bases used in metal finishing are primarily inorganic compounds, they are ionised in water and could be removed by contact with ion exchange resins.

Selectivity of ion exchange resins in order of decreasing preference

Strong acid cation exchanger
$Ba^{2+} > Pb^{2+} > Sr^{2+} > Ca^{2+} > Ni^{2+} > Cd^{2+} > Cu^{2+} > Co^{2+} > Zn^{2+} > Mg^{2+} > Ag^+ > Cs^+ > K^+ > NH_4^+ > H^+$

Strong base anion exchanger
$I^- > NO_3^- > Cl^- > CN^- > HCO_3^- > OH^- > F^- > SO_4^{2-}$

There are synthetic resins that can remove specific ions from the wastewater stream. One such resin cited in literature is iminodiacetic acid [12, 13]. These compounds bond the metal ions into a ring otherwise known as *chelation*. Chelating resins are an order of magnitude more expensive than other resins. The concentrated metal can be recovered to reuse the metal back in the process.

The advantages of ion exchange methods are

- it can operate on demand
- achieves essentially zero level of effluent contamination
- there are a large variety of specific resins available.

The main costs of ion exchange processes are in the cost of regeneration, which is dependent on the type of resin employed, feed-stream quality and the operating arrangement.

The efficiency of ion exchange systems are subject to

- influent quality – higher suspended solids loading increases the pressure drop requiring frequent backwashes and regenerations and leakage of heavy metals
- irreversible fouling of resins due to precipitation of calcium sulphate or other dissolved substances within the resin or adsorption of large organic molecules blocking resin sites
- inlet concentration of the substance being removed
- concentration of regenerant, frequency of regeneration and disposal of spent brine solutions
- variable effluent quality
- resin loss, ageing of resin resulting in loss of active ion exchange sites.

8.9   Adsorption

Adsorption using granular activated carbon (GAC) is a reliable and effective way of removing small quantities of soluble organic impurities in water and wastewater streams particularly those substances that are not readily biodegradable. Activated carbon adsorbs aromatic solvents such as benzene, toluene and xylene; chlorinated aromatics such as chlorobenzene; surfactants such as detergents; textile dyes aromatic and high molecular weight amines; hydrocarbons such as kerosene; pesticides and herbicides; viruses and many other pollutants. They are frequently used as a pretreatment step before RO membrane systems. Wastewater treatment applications include effluent water treatment in the remediation of contaminated ground water, colour removal in textile effluents, car wash facilities and landfill leachate treatment.

Activated carbon is made from bituminous and lignite coal, bone as well as from coconut shells. They all have different properties.

Activated carbon removes organics by adsorbing them onto its microporous surface. The adsorption capability and the rate of adsorption of activated carbon is determined by surface area, pore size, method of manufacture, and characteristics of the solute such as molecular weight, polarity, concentration, pH and temperature. It is available as Powdered Activated carbon (PAC) with a particle size less than 100 μm and GAC which has a particle size of 1–3 mm. GAC is the more common application in industrial wastewater treatment applications. PAC is used as an additive in the activated sludge process by mixing powdered activated carbon into the aeration basin. In general, GAC can remove over 90% of the organics in many industrial wastewaters [14] and usually used when the dissolved organics concentration is less than 10 mg/L [15]. The effluent stream concentration will often be less than < 1 μg/L. However, activated carbon is not effective on organics such as alcohols and aldehydes [15].

GAC is housed in pressure vessels similar to sand filtration.

Problems with GAC units are that the carbon particles can break off and plug RO prefilters. It can also change the characteristics of the water raising the hardness level and in some cases pH. Being softer lignite-based activated carbon in particular is prone to greater attrition. Activated carbon is also an excellent environment for bacterial growth. Chlorine may not be effective in such cases. The disposal of spent GAC is another issue. Some sites have steam regeneration facilities to remove the organics and recover the GAC. Therefore careful consideration needs to be given to these issues before making a decision to install GAC units.

8.10 Membranes for Removal of Dissolved Ions

8.10.1 Overview

Since their advent in the 1960s, membrane processes are the preferred option for the treatment of water and wastewater. Costs of membrane systems have reduced dramatically and, coupled with technological advances in membrane design, membrane options and operating limits, the range of applications in water and wastewater treatment is increasing rapidly.

In membrane filtration, membranes separate the components of a fluid under pressure. The membrane pores, being extremely small, allow the selective passage of solutes. This section will briefly discuss membrane systems.

The popularity of membrane processes arise from the fact that they are effective in the removal of both dissolved and suspended solids. Their advantages are

- ability to recover both the clean water (permeate) and concentrated streams (reject) without chemically modifying them
- applicable to a wide range of processes
- contain relatively few moving parts
- modular construction enables scaling up or down.
- membranes can be custom selected to achieve the desired water quality.
- compact design – means reduced foot print
- short start up times compared to biological systems
- minimal chemical pre-treatment
- low energy requirements.

Their disadvantages are

- upfront capital costs can be significant
- finite membrane life requires periodic membrane replacement
- prone to irreversible fouling requiring more frequent membrane replacement
- strong oxidising chemicals, solvent and extreme pH can degrade certain membrane elements
- limited application at high temperatures
- limited options for the disposal of concentrated brine stream.

This section will be devoted to a brief discussion of pressure-driven membranes such as

- microfiltration (MF)
- ultrafiltration (UF)
- nanofiltration (NF)
- reverse osmosis (RO) membrane applications for water reuse.

Electrodialysis (ED) and electrodialysis reversal (EDR) will only be mentioned briefly.

Figure 8.16 shows filtration chart with the sizes of common particles, membrane types, approximate molecular weight and their ability to remove particles.

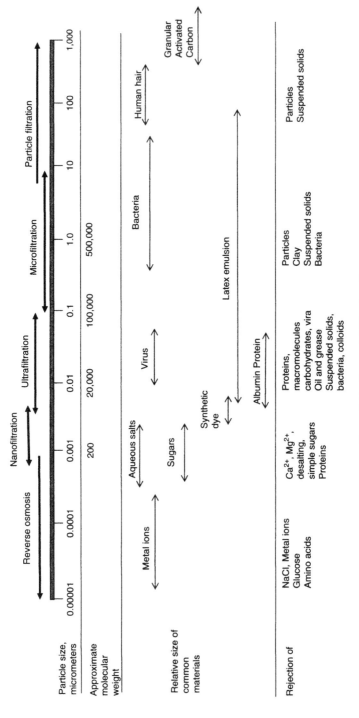

Figure 8.16 Tangential Filtration Chart

8.10.2 Dead-End and Cross-Flow Filtration

Filtration separation is based on the particle size and the volumetric throughput of a filter. Volumetric throughput is proportional to the area and pressure and inversely proportional to the thickness of the filter cake and dynamic viscosity of the fluid.

$$\text{Volumetric throughput} \propto \frac{\text{Area} \times \text{Applied pressure}}{\text{dynamic viscosity} \times \text{thickness of filter cake}}$$

(8.3)

Everything being equal, the volumetric throughput will be a function of the thickness of the filter cake. The two types of membrane filtration are based on this principle. They can be classified as

1. Dead-end filtration
2. Cross-flow filtration.

8.10.2.1 Dead-end Filtration

In conventional dead-end filtration, the flow is perpendicular to the membrane surface. There are only two streams present – feed and permeate streams. The fluid passes through the membrane and all particles larger than the pore sizes of the membrane concentrate at the membrane surface. The trapped particles gradually build up a *filter cake* on the surface of the medium, creating a barrier for the solutes which results in increased resistance to the filtration process. Consequently, dead-end membranes systems are limited to low solids applications or to a few types of applications – such as cartridge filtration of boiler feedwater, microfiltration, or ultrafiltration of municipal treatment [16]. Figure 8.17 shows a graphical illustration of dead-end filtration.

8.10.2.2 Cross-flow Filtration

The limitations of dead-end filtration are overcome in cross-flow filtration. In cross-flow filtration there are three streams. The feed stream flows parallel to the membrane surface (tangential) creating a pressure differential across the membrane surface. Consequently some of the particles pass through the membrane which is known as the *permeate* stream. The rest of the solids flow across the membrane scouring the surface and cleaning it in the process – this stream is known as the *reject* stream. Since the solids are continuously removed it minimises the filter cake thickness to a few microns on the membrane surface and facilitates the continuous operation of the system. Therein lies the advantage of cross-flow filtration to effectively separate micron, submicron, molecular and ionic range particles. Figure 8.17 shows these two different filtration mechanisms.

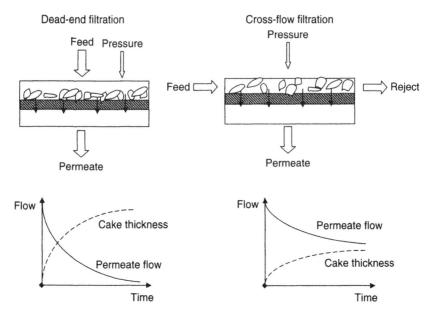

Figure 8.17   Dead-end and cross-flow filtration mechanisms.

8.10.3 *Membrane Types*

There are four basic types of membranes based on pore size and rejection characteristics. These are

1.   Microfiltration (MF)
2.   Ultrafiltration (UF)
3.   Nanofiltration (NF)
4.   Reverse Osmosis (RO).

8.10.3.1 *Microfiltration*

Microfiltration (MF) is the most open media with pore sizes ranging from 0.01 to 1 μm. Microfiltration membranes are used to separate suspended solids, bacteria, algae and cysts and it is capable of being operated in the dead-end and cross-flow filtration modes. Given the relatively large pores, the operating pressure of MF membranes is the lowest of all membrane processes. Typically applied pressure varies from 21 kPa to 345 kPa (3–50 psig). A wide variety of materials can be used to make microfiltration membranes as shown in Table 8.16. Common applications are in the removal of *Giardia, Cryptosporidium*, viruses, as pre-treatment for reverse osmosis membranes, degreasing, metal recovery and in membrane bioreactors (MBR). Figure 8.18 shows a photo of microfiltration membranes.

Table 8.16 Comparison of membranes

	MF	UF	NF	RO
Pore size, μm	0.01–1	0.005–0.01	0.0001–0.001	<0.0001
Molecular weight cutoff (MWCO)	>100,000	1,000–500,000 [17]	300–1,000	100–300
Suspended solids removal	Yes	Yes	Yes	Yes
Micoroorganisms removal	Limited. Protcozoa, cysts, Bacteria and algae	Limited. Protozoa, cysts, Bacteria algae and viruses	All	All
Dissolved Organics removal	None	Yes	Yes	Yes
	Only large compounds >100,000 MW retained.	Insoluble BOD, COD and large organic molecules Macro-molecules, proteins, mono and polysaccharides >2500 MW	Organics >300 MW. Simple sugars and Trihalomethane compounds.	Organics >100 MW
Dissolved inorganics removal	None	None	20–85% rejection	All dissolved salts to 95–99% rejection.
Common applications	Suitable for suspended solids remova . Used as pre-treatment for RO units in treatment of sewage effluent, meta finishing, metal plating and printed circuit board applications.	Reduces turbidity by 99%. Pretreatment for other purification systems. Performs well in oily wastewater, in metal finishing applications, laundry and textile industries.	Has water softening capabilities, i.e. Rejection of divalent salts (Ca and Mg) and some rejection of monovalent salts such NaCl. decolourising. Used to separate sugars	Used as pre-treatment for demineralisation ion exchange and where low dissolved solids effluent is required.

(Continued)

Table 8.16 (*Continued*)

	MF	UF	NF	RO
Disadvantages	Certain materials fouled by oily wastewater. No reduction in dissolved solids.	No reduction in dissolved salts or removal of hardness.	Prone to fouling by colloidal materials and water-treatment polymers. Pretreatment required.	Fouled by colloidal materials, water-treatment polymers and sparingly soluble salts. Pretreatment required. Generally SDI* < 3
Recovery rate	~100%	~75%	~70–85%	~50–85%.
Membrane structure	Polyvinylidene fluoride (PVDF), Polysulfones (PSO), polypropylene, polyacrylonitrile, ceramics	PVDF, PSO, Cellulose acetate, thin film	Cellulose acetate, Thin Film composites	Cellulose acetate, Thin Film composites, polysulfones
Membrane module	Tubular, hollow fiber, plate-and-frame and spiral wound	Tubular, plate-and-frame, spiral wound, hollow fibre	Tubular, plate-and-frame, spiral wound	Tubular, spiral wound, plate-and-frame
Operating Pressure, kPa (psig)	20–345 (3–50)	100–1000 (14.5–145)	338–2028 (50–300)	1,500–6760 (225–1000)
FluxL/m²/hr (USgal/ft²/day)	34–680 (20–401)	34–680 (20–401)	8.5–60 (5.0–35)	10–35 (6–21)
Energy usage kWh/m³	2–20	1–10		

* Silt density index

Figure 8.18 Photo of microfiltration membranes

Courtesy of Pall Corporation.

8.10.3.2 Ultrafiltration

Ultrafiltration (UF) membranes have pore sizes ranging from 0.005 to 0.01 μm. They are usually classified based on their *molecular weight cut-off* (MWCO), which is another method used to categorise membranes. The MWCO is defined as the molecular weight of the smallest molecule, 90% of which is retained by the membrane. The units are expressed in Daltons (grams per mole). The UF membranes have MWCO ranging from 1000 to 500 000 Daltons. For an example, if a membrane has an MWCO of 100 000, then it means that 90% of molecules with a molecular weight of 100 000 will be retained by the membrane. A 100 000-MWCO membrane would have a pore size of 0.01 μm [17].

Given that UF membranes have smaller pores than MF, the operating pressures are higher ranging from 100 to 1000 kPa (14.5–145 psig.).

Common UF wastewater applications include the separation of high molecular weight compounds with a MWCO greater than 1000 such as carbohydrates, proteins, paints and dyes; segregation of oil/water emulsions and separation of heavy metals. The UF is also used as pre-treatment for RO membranes. The UF cannot separate dissolved salts, simple sugars salts, sodium hydroxide or amino acids and consequently no appreciable reduction in conductivity of the permeate stream.

8.10.3.3 Nanofiltration

Nanofiltration (NF) membranes have pore sizes smaller than UF with a MWCO in the range of 150–1000 Daltons. The actual pore size is of little significance for NF and RO membranes. Their solids rejection capabilities are based not on pore size but on selectivity towards ionic charge and molecular weight. Thus, NF is more selective towards divalent ions such as Ca^{2+} and Mg^{2+} and simple and complex sugars. Therefore, these membranes have water-softening capability as well as the ability to reduce total dissolved solids. However, monovalent ions such as Na^+ and Cl^- will pass through more freely. Their rejection rate varies between 0 and 50%. This means that NF has an edge over RO when only partial salt reduction capacity is required.

8.10.3.4   Reverse Osmosis

The RO is the tightest possible membrane process in liquid/liquid separation and therefore produces the highest water quality of any pressure driven membrane process. The RO membranes are classified by percentage rejection of NaCl and ranges from 95 to 99.5%. The MWCO is less than 200 Daltons. The operating pressures are highest in RO membrane applications reaching 6760 kPa (1000 psig) or more and it has also the lowest throughput per unit area (flux).

The four separation processes are summarised in Table 8.16.

8.10.4   Membrane Structure

Membranes can be classified according to their material of construction. Membrane materials can be organic polymers (cellulose acetate, polyamides, polypropylene, polysulfones) or inorganic (e.g. ceramic, zirconia). The membrane material composition determines their ability to be used in harsh wastewater environments. These are

- low or high pH
- high temperature > 50° C
- chemical oxidants – Chlorine
- organic solvents
- abrasive metals.

For instance, most polymeric materials are resistant to moderate pH swings except cellulose acetate (CA) membranes which cannot be operated above pH of 8 nor can they tolerate temperatures above 35° C but can tolerate chlorine residuals. On the other hand, polyamide membranes (PA) are susceptible to chlorine but can operate across a wide pH range. Polysulfone (PSO) have exceptional temperature and pH resistance but are more expensive. Polyvinylidiene fluoride PVDF membranes have the advantage that they can be cleaned with strong acids, caustic soda and bleaches. Ceramic membranes are employed in heavy metal removal applications due to their abrasion resistance but are extremely expensive.

8.10.5 Membrane Configurations

Membranes can be classified according to their configuration which are

- Spiral wound
- Hollow fibre
- Tubular
- Plate and frame.

These are described briefly below.

8.10.5.1 Spiral Wound

Spiral wound membranes dominate the market. Its advantages are its compact design and low price per square area and adequate membrane area per unit volume ($300–1000\,m^2/m^3$). These membranes are commonly used in RO and NF systems.

A spiral wound membrane is constructed from two layers of membrane material separated from each other and are glued to a permeate collector fabric. Plastic mesh is used to form a feedwater channel. The membrane layers, permeate collector and feedwater spacer are rolled around a hollow, perforated centre tube that collects the product water. The membrane module is then inserted into a pressure vessel housing. Typically 6–8 membrane elements are linked inside a single pressure vessel. High-pressure feedwater is directed into one end of the element, the permeate is collected in the permeate channel and flows towards the centre tube, and the concentrate exits at the other end of the element. A conventional single element is approximately 100–200 mm (4–8 in.) in diameter and 1000 mm (40 in.) in length. They contain $7–28\,m^2$ ($80–300\,ft^2$) of membrane area respectively.

Due to their close spacing design, spiral wound membranes are susceptible to fouling and therefore adequate pre-treatment is a prerequisite. Wastewater temperatures are kept below 45° C. Figure 8.19 shows a graphical illustration of a spiral wound membrane module.

8.10.5.2 Hollow Fibre

Hollow-fibre membrane modules incorporate a bundle of hollow-fibre membranes in a single element. Feedwater enters the inside of the hollow fibre, permeates through the tube wall and is collected in a perforated centre tube. Their distinct advantage is having a very high surface area per unit volume ($600–1200\,m^2/m^3$). Given their very small diameter (0.5 mm typical to $10\,\mu m$ [16]), they are prone to blockages and fouling. A good pre-treatment system is a prerequisite and they need to operate at low pressures. The design also allows for backflushing of membranes

Figure 8.20 shows photos of an MF symmetric and asymmetric hollow fibre membranes.

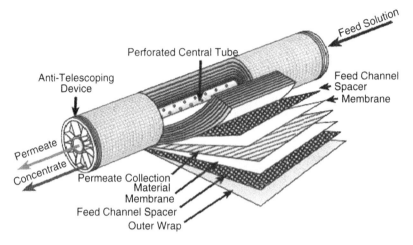

Figure 8.19   A graphical illustration of a spiral wound membrane module
Courtesy of GE Water and Process Technologies.

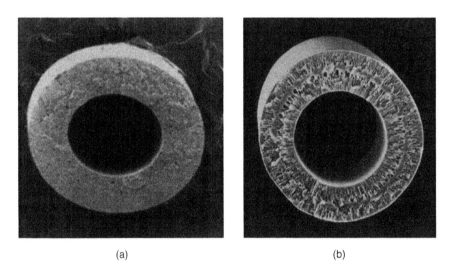

(a)   (b)

Figure 8.20   Photo micrographs of hollow-fibre membranes
Courtesy of Pall Corporation.

8.10.5.3   Tubular

Tubular membranes have a lower surface area than hollow-fibre units ($<100\,m^2/m^3$) and a larger diameter – up to 25 mm (1 in.). The advantage of this type of membrane is that they can operate at higher pressures, are less susceptible to fouling and are more robust than spiral wound membranes. Ceramic membranes are used in this configuration. However, they have several disadvantages. They require more space, and therefore have the highest cost per unit area of all the cylindrical membrane configurations. It is time consuming and cumbersome to change membranes and chemical cleaning costs are high. However, the design allows for backflushing of membranes.

8.10.5.4   Plate and Frame
These modules are rectangular or circular flat sheet systems separated by separators and/or support plates. Cassettes allow membrane modules to be removed individually. Given the low packing density ($100–600\,m^2/m^3$), plate and frame membranes are suitable for specialist high fouling applications. The design allows for backflushing of membranes. Figure 8.21 shows a stack of plate-type membranes.

All membranes suffer from fouling. Physical fouling occurs due primarily to the formation of a boundary layer that builds up naturally on the membranes surface during the filtration process. Traditional cross-flow membranes systems such as spiral wound membranes have very fine clearances. These fine tolerances limit their use to low solids waters or require extensive pretreatment to handle high fouling applications such as landfill leachate. As explained previously cross-flow filtration relies on high velocity fluid flow, pumped across the membranes surface as a means of reducing the boundary layer effect. In cross-flow designs, it is not economic to create shear forces in excess of 15 000 inverse seconds [16, 18–20], thus limiting the use of cross-flow to low-viscosity (watery) fluids. In addition, increased cross-flow velocities result in a significant pressure drop from the inlet (high pressure) to the outlet (lower pressure) end of the device, which leads to premature fouling of the membrane.

Some recent breakthroughs in membrane filtration technology have now made it possible for the treatment of some previously difficult separation applications. New "plate and frame" type membrane modules can tolerate very high levels of TSS, organics and COD. These overcome the limitations of conventional membrane modules and are better able to handle the high fouling and plugging applications like landfill leachate.

Figure 8.21   A photo of a stack of plate-type membranes

Courtesy of Pall Corporation.

To alleviate the limitations on solids entering a membrane system, new open channel type plate and frame membrane modules have been developed. Two leading designs include the VSEP by New Logic Research Inc. and Disc Tube™ Module by Rochem Separation Systems.

1. Vibratory Shear Enhanced Processing (VSEP) system
The Vibratory Shear Enhanced Processing (VSEP) system is one such technology. VSEP system is able to generate high crossflow and turbulence through torsional oscillation, which keeps the feed liquid homogenous and evenly concentrated. The membranes are vibrated in resonance at a frequency of about 53 Hz (times per second) with an amplitude of 3.2 cm (11/4 in.) to the membrane and the membrane displacement is equal to 1.9 cm (3/4 in.) peak to peak at the perimeter imparting a shear force of 120 000–150 000 inverse seconds [16, 18–20]. Figure 8.22 shows a comparison of the flow dynamics. Consequently, VSEP system can be used in high fouling and high dissolved solid applications such as in black liquor recovery (pulp mills), desalter effluent (oil refining), oil/water emulsions, mine water and landfill leachate with minimum pretreatment. New Logic claims the ability to concentrate the reject stream up to 50% solids (500 000 mg/L). The VSEP system also tolerates high temperatures up to 70–90° C.

Since VSEP is not limited by solubility of minerals or by presence of suspended solids, it can be used to treat the RO reject of conventional spiral wound membranes.

The industrial VSEP machine contains many sheets of membrane, which are arrayed as parallel disks separated by gaskets. The disk stack is contained within a Fiberglass Reinforced Plastic (FRP) cylinder. A 185 m² (2000 ft²) membrane is contained in one VSEP module with a footprint of only 100 × 100 (4 ft × 4 ft) [19]. The result is that the horizontal footprint is very small. Figure 8.23 shows VSEP system.

Case Study: Removal of Calcium Carbonate ($CaCO_3$) Slurry water using a VSEP microfiltration system [19]

The customer required equipment to dewater $CaCO_3$ slurry as part of a manufacturing process. Pilot testing of spiral wound membranes, centrifugal separators and rotary drum filters failed, due to technical limits, high capital and operating costs. The VSEP system was trialled where a hot 7% solution of $CaCO_3$ is fed into a VSEP at 3.8 m³/hr (17 US gpm). The concentrated stream is at 30% solids. A second pass unit concentrates this to 50% – solids approaching the limits of 'pumpability'.

The biggest advantage of the VSEP in this application was energy savings, reducing pre-treatment equipment costs and labour costs. The system has a payback period of less than one year.

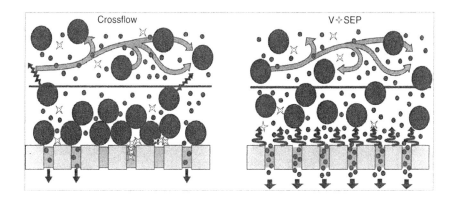

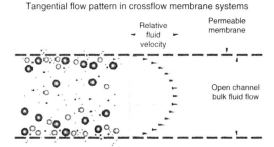

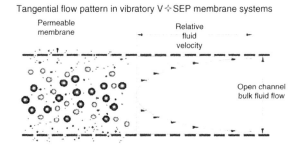

Figure 8.22   A fluid dynamics comparison between VSEP and conventional cross-flow filtration

Courtesy of New Logic Research Inc.

2. Disc Tube technology

Pall Rochem's Disc Tube™ Module (DTM) technology was first developed in Germany and contains multiple leaf layers of membranes stacked in a column with spacer setting the gap between them. The Disc Tube™ modules are a tangential flow separation system. This tangential flushing action and the optimised hydrodynamics in the module are the primary factors in keeping the membrane clean and operating properly, which means that the modules rely on high turbulence and high cross-flow to keep the membrane surface clear of suspended solids cakes and other formations that would blind

Figure 8.23   A photo of VSEP system showing the small footprint

Courtesy of New Logic Research Inc.

Figure 8.24   Photo of Pall Rochem Disc Tube™ module

Courtesy of Pall Corporation.

the membrane. Unlike VSEP these do not have the vibration capabilities. Figure 8.24 shows a photo of Disc Tube™ module.

Membrane systems can be arranged in a number of ways. The simplest design consists of a feed tank, a feed pump, membrane system and tanks to hold the permeate flow and the reject stream. The permeate quality may be adequate to the need or may require another set or a different type

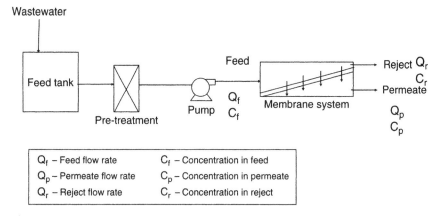

Figure 8.25   Schematic of a single stage membrane plant

of membrane to reach the desired water quality. Cleaning chemicals and sanitisers also require holding tanks and ancillary piping for regular cleaning of membranes.

A simplified schematic of a membrane system is shown in Figure 8.25.

8.10.6   Membrane Performance Monitoring

Membranes are expensive to replace and therefore lengthening membrane life reduces operating costs. The performance of membranes decline with time and therefore membrane performance monitoring helps to detect product quality, permeate flow and onset of membrane failure.

Membrane fouling is one of the main reasons for premature membrane replacement and product quality issues. Membrane fouling results in decreased product quality, reduced flux and an increase in pressure drop across the membrane. Fouling can be either reversible or irreversible. Fouling can be due to inherent system conditions, inadequate pretreatment and/or improper operation.

Fouling can occur due to

- physical blockage of the membrane surface
- chemical scaling of membrane surface, due to a mechanism known as *concentration polarisation*, when salts precipitate on the membrane surface.
- biofilm formation on membrane surface
- *solute absorption*, where adsorption of small molecules makes the pores smaller. For example, in ultrafiltration membranes this is a common cause for a decline in performance, when protein molecules get absorbed into pore walls.

There are other reasons for membrane failure such as: thermal shocks, operating at low or high pH, membrane compaction or improper use of

Table 8.17 Common failure mechanisms

Type of failure	Cause	Remedial measure
Fouling – physical blockage	Inadequate pre-treatment.	Install cartridge filtration, MF/UF before RO. Back flushing, back pulsing, increasing turbulence and cleaning generally restores originally capacity. Cleaners are acids, alkalis, enzymes and detergents.
Fouling – precipitation of sparingly soluble salts	Inadequate pre-treatment. Too high permeate recovery can lead to precipitation of sparingly soluble salts. More of a problem in NF and RO plants	Reduce pH; add ion exchange upstream of membrane plant; inject chemical inhibitors; reduce permeate recovery rates. Can result in irreversible decline in performance.
Fouling – microbiological	Biofilm formation due to inadequate disinfection during operation and/or poor storage practices.	Disinfect with approved products. Use recommended storage practices. Generally reversible except in the case of cellulose acetate (CA) membranes.
Fouling – adsorption of macromolecules in to the membrane	Inadequate pre-treatment. Certain surface active macro-molecules or iron can cause irreversible blockage.	Pre-treatment, eliminate cationic coagulants if it is the cause and or use a different membrane material which is less prone to the particular fouling type.
Membrane degradation due to chemicals.	Operations at high or low pH. Oxidants and organic solvents can damage membranes.	Follow manufacturer's recommendations. Use an appropriate membrane material. Dechlorinate.
Mechanical failure	Hydraulic shocks, high temperature, drying out of membranes due to poor storage practices.	Follow manufacturer's recommendations. Use an appropriate membrane material.

oxidants. The common causes and their remedial measures are listed in Table 8.17.

Silt density index is a filtration test to determine pretreatment requirements to guard against physical fouling of membranes. Scaling gives the potential for chemical fouling and Flux determines the product flow rate per unit area of membrane surface.

8.10.6.1   Silt Density Index

The Silt Density Index (SDI) is a simple filtration test, which measures the fouling tendency of the water. The SDI indicates the rate at which membrane's pores plug. ASTM D4189 describes the procedure. An SDI of 0 is very clean water. An SDI of 6.67 is dirty water. Most equipment vendors of RO and NF equipment specify SDI to be less than 5. However, for conventional spiral wound RO membrane systems, SDI of less than 3 is preferred to minimise problems caused by suspended solids blocking the brine spacers in a RO membrane module. Waters of higher SDI values needs to be filtered using multimedia filters followed by $5\,\mu$ cartridge filters, chemical treatment, or using MF/UF prior to RO/NF membranes. The MF membranes will guarantee an SDI of less than 3 and UF usually achieves SDI of below 2 and frequently below 1. In operation, any increase in SDI needs to be investigated because it means that either the feed quality has changed or a there is a failure in the pre-treatment equipment.

8.10.6.2 Assessment of Scaling Tendencies

Scaling is more of a problem in NF and RO membranes because these membranes convert 75–90% of the feedwater into product water and in the process they reject dissolved ions. As discussed previously, as RO membranes produce pure water, the feed water becomes concentrated. Some of the chemical compounds that are soluble in the RO feed water may no longer be soluble. More importantly, at the membrane surface the concentration of the mineral ions can be several fold higher than what is in the bulk feed water due to a phenomenon known as *concentration polarisation*. The higher the permeate recovery, the greater the concentration of the minerals such as Ca^{2+} in the reject stream which can lead to scaling of the membrane surface resulting in increased energy consumption and chemical cleaning frequency. Therefore, it is important to determine the solubility potential of sparingly soluble compounds at the membrane surface. The Langelier Saturation Index (LSI) or the Stiff Davis Saturation Index (SDSI) is used to evaluate scaling tendencies of calcium carbonate. Chapter 5 gives a description of LSI. The SDSI is a variation of LSI. When the TDS is greater than 5000 mg/L, SDSI is recommended to be used. [5] It is important to maintain a negative LSI of the RO concentrate to ensure no scaling on the membrane surface occurs. This is achieved by acid addition. Use of a weak acid cation exchanger can also be used to remove carbonate hardness, followed by a degassing unit to remove CO_2.

8.10.6.3 Membrane flux

Generally, there is a correlation between pore size and flow rate. The smaller the pore size (more accurately membrane resistance) the lower the flow rate per unit area of membrane – otherwise known as *flux*. It is expressed as m^3/m^2 s (SI units). The more common units are L/m^2/hr (LMH) or gal.ft^2 day (GFD). For a given membrane and feed water quality, flux is a function of applied pressure and water temperature. Generally, the better the feed water quality, the higher the flux. Flux increases with temperature and pressure. However higher the temperature the higher the salt passage through the membrane.

The higher the design flux rate the lower the capital cost and size of the membrane plant. Pretreatment of RO membranes with MF or UF membranes typically increases the flux rate of RO elements.

8.10.6.4 Permeate Recovery

Permeate recovery is defined as the ratio of permeate flow (Q_p) to feed flow (Q_f), usually expressed as a percentage. This is shown in Equation (8.4). It is an expression used to describe how efficiently the system is being operated, and is also used to determine the extent of concentration of the fluid solutes.

$$\text{Permeate recovery } (R\%) = \frac{Q_p}{Q_f} \times 100 \qquad (8.4)$$

From formula 8.4 it follows that, the higher the permeate recovery, the higher the permeate flow, the less the reject flow rate. On the other hand, the higher the permeate recovery, the poorer the permeate quality and the more concentrated the reject stream – increasing scaling potential of the membrane surface. Typically, for naturally occurring surface waters, permeate recovery more than 75–80% is not recommended even with good pretreatment. A 5% increase in permeate recovery means that at 75% the feed water concentrates four times at the concentrate outlet and at 80% recovery it concentrates 5 times. The permeate quality will decrease by approximately 25% as permeate recovery moves from 75 to 80%. For wastewater reuse permeate recovery is normally around 70–80%. Reject streams can serve as feed for subsequent modules and this arrangement is known as 'reject staging' and produces higher recovery rates than the single pass arrangement.

The concentration in the reject stream can be approximated by multiplying the feed concentration by the concentration factor (X) as shown in formula 12.

$$C_r = \frac{C_f}{(1 - R\%)} = XC_f \qquad (8.5)$$

where

C_f, C_r given in concentration in mg/L

Worked example

If silica concentration in the feed is 20 ppm, what is the concentration of the reject stream at R = 50, 70, 80 and 90%? What is the desirable R%? Refer to Table 8.18

At 80% recovery the concentration factor X is five times. That means, the concentration at the membrane surface is five times the feed concentration. Increasing permeate recovery rates reduce permeate quality. A 5% increase in permeate recovery from 80–85% increases the concentration in the reject stream by 33%. Silica is typically controlled in the range below 100 mg/L, the permeate recovery R is set at a maximum of 80% (higher silica rejection levels can be achieved by increasing the pH > 10).

Table 8.18 Concentration factor vs System Recovery

Recovery Rate (R%)	Concentration Factor (X)	Permeate Flow (%)	Reject Flow (%)	Silica concentration in the reject stream (mg/L)
0 (Feed water)	1	0	100	20
50	2	50	50	40
70	3.3	70	30	67
75	4.0	75	25	80
80	5.0	80	20	100
85	6.67	85	15	133
90	10	90	10	200

8.10.7 Disposal of Brine Streams

Membrane plants produce a reject stream which needs to be disposed of. The common disposal options are to:

  (i)   discharge to the public utility's sewer if the concentrations are acceptable to water utility.

  (ii) To discharge to the sea or other external water source if approval is granted. Increases transport costs if located inland.

  (iii) Concentrate the reject stream further using another membrane system

  (iv) Evaporate the water to dryness in sludge drying beds

  (v) Evaporate the water and recover more distillate using mechanical vapour compression technologies, brine concentration and crystallisers. These options consume power and requires extra capital.

  (vi) Recover saleable materials.

8.10.8 Considerations When Selecting Membrane Systems

Membrane systems require upfront financial costs. The penalty for not getting it right is prohibitive. Quite often the unforeseen problems occur at the time of commissioning but by then it is too late (and costly) to change the design and the blame game starts between the vendor and the customer.

Therefore, before purchasing a membrane system the more detailed the analysis is, the lower the risk. It is always preferable to do on-site plant trials – even though the application may be quite common. Spending 10% of the capital costs on a pilot trial will minimise the risk of not getting it right later on.

In selecting and designing a membrane system the considerations include the following.

  1.   What permeate water quality is desired?

  2.   What is the feed water quality?

3. What is the variability of the feed stream in terms of quality, flow, volume and temperature?
4. What pre-treatment systems are in place for removal of suspended, colloidal, microbiological and dissolved substances such as clarification, activated carbon units, chlorination/ozonation, cartridge filters and/or water softeners for removal of temporary hardness?
5. What type of membrane system best suits the need?
6. What strategies are used to reduce LSI if NF or RO membranes are used and how will this affect downstream equipment such as demineralisers? (If acid injection is used, care needs to be taken not to overload the anion exchanger with high CO_2).
7. Is there a valuable resource to be recovered from the process stream?
8. What are the design flux rates?
9. What is the permeate recovery? What penalties are there for increasing permeate recovery rates? Will a second pass improve recovery rates? What are the financial costs of this option?
10. What is the per cent salt rejection?
11. How is the reject stream disposed of? What are the costs?
12. Can the reject stream concentrated further to reduce disposal costs?
13. What regulatory approvals are required for the use of the permeate and discharge of reject streams?
14. What is the expected membrane life and replacement frequency?
15. What are the operating costs for membrane replacement, energy, cleaning, license charges and labour costs?
16. How are the membrane systems integrated into the existing plant?

The capital cost of a membrane system is dependent on the size of the system. It can be approximated as follows [21, 22]:

- pumps   30%
- replaceable membranes   20%
- membrane modules (housings)   10%
- pipework, valves, framework   20%
- control system   15%
- other   5%

The operating costs for a membrane system consist of [21]:

- membrane replacement   35–50%
- cleaning   12–35%
- energy   15–20%
- maintenance labour   15–18%

8.10.9 *Electrodialysis and Electrodialysis Reversal*

Electrodialysis (ED) is an electrochemical membrane process that involves the movement of ions through anion and cation selective membranes from a less concentrated solution to a more concentrated solution by the application of a direct current (DC). The difference in these membrane processes is that no pressure is applied. Direct current causes the charged ions to move towards the anode (+) and cations towards the cathode (−). Ion-selective semipermeable membranes placed in between the electrodes alternatively allow the passage of only hydrogen and hydroxyl anions to pass through to the respective electrodes. In the process the impurities are trapped within the membranes and as a result produce a concentrate stream. In the ED process, with time, the membrane process becomes saturated with charged ions reducing the recovery rate.

The Electrodialysis Reversal (EDR) process overcomes this problem by reversing the polarity of the electrodes every 15 minutes. Polarity reversal causes the concentrating and diluting flow streams to switch after every cycle. This results in cleaning of the membranes, by sending high-quality water in the compartment that was previously filled with the reject stream.

The ED/EDR processes is suitable for desalting brackish water with TDS feedwater concentrations of up to 4000 mg/L. After 4000 mg/L the electricity costs start to increase dramatically. It is not suitable for the removal of organics. In wastewater treatment, applications include concentration of RO reject streams, mining water reuse and cooling tower blowdown treatment.

References

[1] Goto T. *Industrial Water Reuse in Japan*. Vol. 5/3, Desalination & Water Reuse. 1992.
[2] US EPA Water Reuse Manual. Washington D.C. 1998.
[3] Goronszy M.C., Eckenfelder W. and Froelich E. *A Guide to Industrial Pretreatment*. Chemical Engineering. McGraw Hill. November 1992.
[4] Arterburn, R.A. *The Sizing and Selection of Hydrocyclones*. Krebs Engineers. Menlo Park. CA.
[5] Ford D. and Tischer L.F., *Industrial Waste*. pp 20–25. July/August 1977.
[6] Capps R.W., Matelli G.N. and Bradford M.L. *Reduce Oil and Grease Content in Wastewater*. Hydrocarbon processing. June 1993.
[7] New Logic Research Inc. *Using Vibrating Membranes to treat oily wastewater from a waste hauling facility*. www.vsep.com.
[8] Davies P.S. *The Biological Basis of Wastewater Treatment*. Stratkelvin Instruments Ltd. Glasgow. 2005.
[9] Environmental Technology Best Practice Programme. GG 156 – *Cost-Effective Effluent Treatment in Paper and Board Mills*. UK. 1999.
[10] Spanjers H. *Anaerobic Treatment of Textile Wastewater*. Lettinga Associates Foundation. Wagenigen, The Netheralnds.

[11]   Water Environment Federation. *Pretreatment of Industrial Wastes.* Manual of Practice FD-3. Alexandria, Virginia. 1994.

[12]   United States Environmental Protection Agency. Office of Research and Development. Capsule Report – *Aqueous Mercury Treatment.* Washington D.C. July 1997.

[13]   Davis F.S. *Heavy-Metals Removal Processes in Industrial Waste Streams.* Industrial Water Treatment. pp 38–41. May/June 1994.

[14]   Eckenfelder Jr. W.W., Patoczka J. and Watkin A. *Wastewater Treatment.* Chemical Engineering. September 2 1985.

[15]   Environmental Technology Best Practice Programme. GG 37 – *Cost-Effective Separation Technologies for Minimising Wastes and Effluents.* Hartford, UK. 1996.

[16]   Judd S. and Jefferson B. *Membranes for Industrial Wastewater Recovery and Re-use.* Elsevier Advanced Technology UK. 2003.

[17]   Von Gottberg A.J.M. and Pereschino J. *Using Membrane Filtration as Pretreatment for Reverse Osmosis to Improve System Performance.* Ionics Incorporated. Proceedings of the North American Biennial Conference of the American Desalting Association, 7 August 2000.

[18]   Reynolds G. and Glod R. *An Examination of the Use of Vibrational Shear in Ultra-Filtration, Nano-Filtration and Reverse Osmosis Membrane Water Treatment.* International Water Conference. Pittsburgh, Pennsylvania. 22–26 October 2000.

[19]   Bian R., Yamamoto K. and Watanabe Y. *The Effect of Shear Rate on Controlling the Concentration Polarisation and Membrane Fouling.* Proceedings of the Conference on Membranes in Drinking and Industrial Water Production, Vol. 1, pp. 421–432, Desalination Publications, L'Aquila Italy. October 2000.

[20]   New Logic Research Inc. *Black Liquor Treatment for Pulp Mills.* www.vsep.com.

[21]   Byrne W. *Reverse Osmosis – A Practical Guide for Industrial Users,* Tall Oaks Publishing Inc, Littleton, USA. 1995.

[22]   Environmental Technology Best Practice Programme. GG 54 – *Cost Effective Membrane Technologies For Minimising Wastes and Effluents.* UK. March 1997.

Chapter 9

Making a Financially Sound Business Case

9.1 Introduction

It is not uncommon for good projects that could save water – to sit on the shelf. The main reason for this is that the business case fails to meet the organisation's financial criteria. Not many businesses can function without water. However there is a mismatch between the reluctance to pay for the resource and the scarcity of the resource. Consequently to obtain funds, the project's business case needs to demonstrate how the project adds to shareholder value in addition to environmental value. In other words the challenge is to show that *saving water makes good business sense*. Any cost savings from a project flows directly to the bottom line as shown in formula (9.1).

In a manufacturing plant the formula could be expressed as

$$\text{Shareholder value added} = (\text{Water associated costs \$/unit} \\ - \text{Water associated costs after improvement \$/unit} \qquad (9.1) \\ \times \text{production capacity}.$$

In other industries the derivation of the formula is not that straightforward.

To obtain funding the business case needs to demonstrate how the project.

- improves the bottom line
- aligns with corporate strategy
- increases shareholder value
- decreases financial, environmental, health and safety or other regulatory risk
- increases the competitive advantage of the organisation by being able to command a premium price for the product or attract more sales
- increases the reputation of the business unit or organisation both internally and externally.

This is especially so when companies are struggling to achieve a return in their core business – due to effects such as

- poor sales
- reduced margins
- global competition
- reduced tariffs
- business confidence in the economy decreases
- high oil prices
- high interest rates
- recession.

In such instances, costs are scrutinised more closely and only core business activities tend to get funded, especially in industries such as bulk chemicals, steel, food, oil, paper and other commodity products, where the end products cannot be easily differentiated from a competitor's products. By the same token these industries have a high reliance on community goodwill.

This chapter explores how one can present a financially sound business case so that projects are implemented – and do not gather dust.

9.2   Management Functions

Financial hurdles are a measure of management performance. Senior management is responsible for increasing shareholder value whether an organisation is public or private. Upper management is also responsible for reducing risks. Therefore the business case in order to be successful needs to meet the performance objectives and priorities of both middle and senior level management. Table 9.1 shows how the level of importance of key decision criteria, rate managers according to seniority within the organisation [1].

9.3   Making a Good Business Case

The task of the water conservation manager, in developing an attractive business case, is to ensure that all costs, savings and risks are identified especially when the project has a high capital expenditure component.

Table 9.1 Degree of importance* of key decision criteria, according to position in the company

Decision Factors Importance Level	Operations Managers and their Report	Middle Management	Chief Operating, Executive and Financial Officers
Product Features, Functionality	$$$	$$	$
Cost savings	$$	$$$	$$$
Return on capital	$	$$	$$$
Reducing Risk	$$	$$	$$$
Image	$	$	$$$

* Degree of importance – Low – $; Medium $$; Very important – $$$.

Very often the tangible (and intangible) costs and benefits of the water-saving project are not adequately identified. Externalities (parameters outside the organisation) need to be factored in. A pollution tax or future water prices are just two examples. Future expansion plans for the site and its impacts on the organisation will also need to be included in the equation. This relates to security of water supply.

With water scarcities potentially impacting on many countries, there is the real threat of governments rewarding those organisations that use water resources more efficiently than others. In other words, companies' expansion plans may be curtailed until they can demonstrate how water- and energy-efficient they are. These opportunity costs need to be considered and factored into the equation.

Sometimes the benefits are totally unrelated to the operations department but may reside in the improving the company's image (sales), increasing the profile amongst customers (sales), lead to a reduction in transport costs or improving labour relations (human resource department).

The business case should also identify the benefits/risks of not implementing the project as well as the associated benefits/risks of implementing the project – and rank them on a scale of 1 to 10 or use a standard risk matrix as shown in Table 9.2.

The matrix needs to include technical, financial, environmental and other business risks. Sensitivity analyses will highlight if the project is overly sensitive to interest rates. Risk assessment is discussed further in a later section.

The project benefits need to be related to metrics that the organisation uses for assessing performance (such as cost of goods sold (COGS) per unit of output) and answer the question – does the project improve COGs?

The financial analysis of water conservation projects is dependent upon having accurate information about operating costs. Few companies, however, possess good process-focussed information. In most manufacturing operations, while direct raw material and direct labour costs are tracked closely, many organisations lack systems to track and allocate "indirect" costs

Table 9.2 Risk matrix

Consequences	Likelihood				
Seriousness of event	Likelihood of occurrence				
	Very likely	Likely	Occasional	Unlikely	Very Unlikely
EXTREME	HIGH	HIGH	HIGH	MEDIUM	MEDIUM
MAJOR	HIGH	HIGH	MEDIUM	MEDIUM	LOW
MODERATE	HIGH	MEDIUM	MEDIUM	LOW	NEGLIGIBLE
MINOR	MEDIUM	MEDIUM	LOW	NEGLIGIBLE	NEGLIGIBLE

Table 9.3 Conventional cost accounting

Cost Pools	$ (in 1,000s)
Direct material cost (raw materials that become part of the finished product)	
(A) Inventory difference	11,000
(B) Purchases of direct materials	73,000
Cost of direct materials available for use C = (A+B)	84,000
Direct manufacturing labour costs (D)	17,750
Manufacturing overhead costs (E)	
Indirect labour (work that supports production)	4,000
Indirect supplies (materials not part of the product)	1,000
Utilities (heat, light, power and water)	1,750
Facility costs	500
Depreciation (plant building and equipment)	1,500
Miscellaneous	500
(E)	9,250
Manufacturing costs incurred (F) = C + D + E	111,000
Corporate expenses (administration, marketing and sales)	3,000

(by process or product) and therefore such costs are not readily accessible or are often termed as "fixed costs".

Project benefits need to be related to metrics that the organisation uses for assessing performance such as cost of goods sold per unit output (COGS/unit).

Table 9.3 shows a schedule for cost of goods manufactured under a typical conventional accounting method.

9.3.1   Identifying Hidden Costs

As shown in Table 9.3 it is easy to identify direct costs. The hidden costs, however, are not shown in accounting schedules and these must be identified using a variety of means such as

- measurement
- interviews with operators (for an example, to gauge the amount of time spent in cleaning)
- compiling billing data from water bills to disaggregate water-related charges to fixed and variable costs
- EPA fines for non-compliance and so on.

Examples are given below

Regulatory

- reporting
- monitoring/testing

- inspections
- non-compliance fines for discharges to the environment
- insurance costs.

Conventional production costs

- Cost of raw materials wasted
- Cost of water
- Wastewater discharges
- Internal treatment
- Replacement of equipment
- Trade waste charges
- Monitoring
- Pre-treatment before discharge to the sewer
- Transportation of sludge
- Landfill costs
- Labour charges
- Electricity and gas charges
- Cost of steam
- Maintenance costs.

Project implementation costs

- Design and construction costs
- Site preparation costs
- Management and administration costs
- Utility connection costs
- Plumbing costs
- Electrical costs
- Training and commissioning costs.

Voluntary costs

- R&D costs
- Environmental studies and reports
- Habitat protection
- Training
- Audits
- Community sponsorship programmes.

These costs need to be collected for the existing plant in addition to the savings and other costs that will be accrued with the implementation of the water conservation project. This will identify the incremental costs for the project.

Once these costs are identified then cash flows can be calculated as explained in Section 9.4.

Case Study: Identifying the *True Cost of Water* at a food manufacturing plant

A food manufacturing plant located in Sydney consuming $1000\,m^3$/day of water had calculated the *true cost of water* as follows:

	Annual cost
Water usage charges at $1.20/m^3$	$420 000
Sewer usage charges at $1.19/m^3$	$416 500
Trade waste charges for BOD, oil and grease and suspended solids	$138 000
Chemicals for wastewater treatment plant ferric chloride	$55 000
Sludge removal	$312 000
Bacterial culture	$20 000
Deodoriser	$40 000
Trade waste testing	$21 000
Labour charges to operate treatment plant	$45 000
Electricity charges	$100 000
Depreciation expense on pre-treatment equipment	$60 000
Total cost of water	$1 627 500
Total cost of water/m^3	$4.65
If *only* water, sewer usage charges and trade waste were considered, cost of water would only have been/m^3	$2.78

9.4   Computing Cash Flows

Cash flow calculations are important, since it shows the true picture of expenditure or savings throughout the economic lifetime of the project.

Cash flows show the expenditure (outflows) and avoided costs (inflows) of money incurred and generated by a project. Without an understanding of cash flows it is difficult to understand the profitability of a project. Whilst the majority of projects cash inflows mirror the avoided costs that the organisation is saving due to the project, it is important to recognise other cash inflows. Examples included production capacity increases, reduction in marketing costs or on-selling of by-products. A simple example includes a food manufacturing company that was able to sell their sludge from the DAF plant at a higher value ever since they stopped the dosing of polymeric coagulants.

Cash flows can be separated into

- One-time cash flows
- Operating cash flows
- Working-capital cash flows.

One-time cash flows are those that are required to be expended only once – for instance, capital equipment.

Operating cash flow is the expenditure required to keep the plant running.

Working-capital cash flows are required to purchase raw materials until goods are sold and cash deposited in the bank.

The project needs to be assessed over its entire *economic lifetime*. This is defined as the period of time over which a project is expected to add economic value to a business. Typically, economic lifetimes are between 3 and 10 years. This is not to be confused with the *physical lifetime* of the equipment – which is the useful lifetime of the equipment. The economic lifetime is less than the physical lifetime of the equipment.

Any expected inflationary trends and price increases need to be included in costs and savings.

Once all of the costs are known it is useful to arrange them in a timeline over the lifetime of a project – as shown in the case study below.

Case Study: Computing Cash Flows

A cereal company's water conservation project consisted of the installation of a membrane plant at a cost of $1M. The installation costs were $300 000. As a result of the membrane plant the plant can reuse 75% of its incoming water – currently at 1000 m³/day. The true cost of water has been calculated at $2.40/m³. Operation and maintenance costs are $210 000/yr. The economic lifetime of the project is 5 years and inflation is assumed to be 3%. The cost of capital is 20%. The cash flows over its economic lifetime are shown in Figure 9.1.

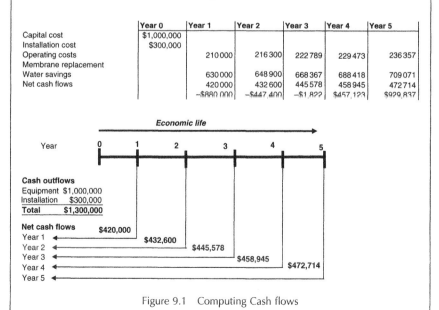

	Year 0	Year 1	Year 2	Year 3	Year 4	Year 5
Capital cost	$1,000,000					
Installation cost	$300,000					
Operating costs		210 000	216 300	222 789	229 473	236 357
Membrane replacement						
Water savings		630 000	648 900	668 367	688 418	709 071
Net cash flows		420 000	432 600	445 578	458 945	472 714
		–$880 000	–$447 400	–$1 822	$457 123	$929 837

Figure 9.1   Computing Cash flows

It must be noted that if depreciation and after-tax cash flows are considered, then a more accurate financial picture of the project is gained. For this it is important to know the depreciation schedules to be applied for the equipment and the marginal tax rates. By depreciating the initial capital equipment over the allowable time horizon, taxable income reduces – and the after-tax cash flows correspondingly increase.

9.5   Investment Appraisal Methods

One of the most important steps in the capital budgeting cycle is working out if the benefits of investing large capital sums outweigh the costs of these investments. The payback period is the most popular method. Traditionally a payback period of 2 years or less is commonly quoted. Many organisations do not take a longer term view unless it is an investment in core business activities or needs to be done due to compliance requirements. Whilst the payback method is a quick and easy way of assessing the project profitability it does not take into account *the time value of money* and *the cash flows that occur after the initial investment is recouped.*

There are **four** common measures of profitability. These can be classified as "Simple" and "Discount Counted Cash flow" methods and are shown in Table 9.4.

9.5.1   Payback Method

When deciding between two or more competing projects, the usual decision is to accept the one with the shortest payback.

Payback is often used as a "first screening method". By this, we mean that when a capital investment project is being considered, the first question to ask is 'How long will it take to pay back its cost?' The company might have a target payback period, and so it would reject a project unless it's payback period was less than the target.

The payback period measures the length of time taken for the return on an investment exactly to equal the amount originally invested. If a number of options are considered, then these are ranked from the shortest to longest payback period. The payback period is calculated as follows.

Table 9.4 Measures of profitability

Measure	Method
Simple	Payback
	Return on investment
Discounted Cash flow	Net present value
	Internal Rate of Return

Initial capital investment: $ Initial Investment

Pre-tax cash flows (CF) for years 1 to $n = \Sigma$ ($CF_{Y1} + CF_{Y2}$

$$+ CF_{Y3} + \cdots + CF_{Yn}) \qquad (9.2)$$

Payback period is equal to the year when

$$\$ \text{ Initial Investment} = \Sigma \ (CF_{Y1} + CF_{Y2} + CF_{Y3} + \cdots + CF_{Yn}) \qquad (9.3)$$

Worked example

A project has an initial investment of $10,000 and the cash inflows are given below:

Years	0	1	2	3	4
Cash inflows/outflows	$-10,000$	2,500	5,000	4,000	2,500

The payback period lies between year 2 and 3.
The sum of money recovered at the end of Year 2 $= \$7,500$
The sum of money to be recovered by the end of 3rd year $= \$10,000$
$-7,500 = \$2,500$
Therefore payback period $= [2 + 2,500/4,000] = 2.6$ years

Or if the annual cash flows are constant, then equation (9.3) simplifies to:

$$\text{Payback period} = \frac{\$ \text{ Initial Investment}}{CF} \qquad (9.4)$$

The advantages of the payback method are

- It is popular; involves quick simple calculation and an easily understable concept.
- It provides an important summary method. How quickly will the initial investment be recouped? Short payback means investors are rewarded quickly
- It is a particularly useful approach for ranking projects under tight liquidity conditions.
- It is appropriate in situations where risky investments are made in uncertain markets that are subject to fast design and product changes or where future cash flows are particularly difficult to predict.
- The method is often used in conjunction with NPV or IRR method and acts as a first screening device to identify projects which are worthy of further investigation.

There are several disadvantages of using the payback period method. These include the following:

- It lacks objectivity. Who decides the length of the optimal payback period? No one does – decided by comparing one investment opportunity against another.
- It ignores the timing of cash flows. Cash flows post payback are ignored.
- It ignores the time value of money. This means that it does not take into account the fact that $1 today is worth more than $1 in 1 year's time. An investor who has $1 today can either consume it immediately or alternatively can invest it at the prevailing interest rate, say 30%, to get a return of $1.30 in a year's time.
- It is unable to distinguish between projects with the same payback period.
- The focus is on cash flows rather than overall business profitability. It may lead to excessive investment in short-term projects.

9.5.2   Return on Investment Method

The ROI method is also called the Return on Capital Employed (ROCE) or the Average Rate of Return. It expresses the profits arising from a project as percentage of the initial capital cost. If it exceeds a target rate of return, the project will be undertaken.

However the definition of profits and capital cost are different depending on what is included. For instance, the profits may be taken to include depreciation. For our purposes ROI is defined as:

$$\text{ROI} = (\text{Annual CF Savings/Initial capital cost}) \times 100 \qquad (9.5)$$

For an example the initial investment is $200,000 and the savings every year are $20,000. Then the ROI is:

ROI = $20\,000/$200\,000 \times 100 = 10%
Payback period = 10 years

If the ROI is greater than the target rate, the project is accepted. The ROI has similar advantages and disadvantages highlighted above for the payback period.

9.5.3   Discounted Cash Flow Methods

The most accurate methods to calculate project profitability are Net Present Value (NPV) and Internal Rate of Return (IRR) methods since both methods takes the *time value of money* into account. They rely on the concept of opportunity cost to place a value on cash inflows arising from a capital investment. The opportunity cost takes into account the interest charges on the capital expenditure, otherwise known as the cost of capital, future value of a project and future cash flows. The future cash flows are discounted by the cost of capital to bring them to present values. Put simply these methods recognise the fact that a dollar earned today is better than earning a dollar tomorrow.

$$\text{Future Value (FV)} = PV \times (1+r)^n \qquad\qquad (9.6)$$

$$PV = FV/(1+r)^n \qquad\qquad (9.7)$$

where

PV – Present Value
R – interest rates
N – time period in which interest is earned.

For an example, if interest charged is 10%, the income that is required in Year 3 to equal to $1000 dollars earned today is

$$FV = \$1000/(1+0.1)^3 = \$1\,331$$

It is customary to consider the future cash flows in present day dollars by considering the PV of those cash inflows.

The interest assigned to a project is the organisation's *cost of capital*. The cost of capital is different from the prevailing bank interest charges since it reflects the organisation's capital structure. In simple terms the cost of capital takes into account the rate of interest on outstanding debt, the opportunity cost of money (prevailing bank interest rates for investments) and the relative risk of the investment.

9.5.4 Net Present Value Analyses

Net Present Value (NPV) is equal to the present value of future returns, discounted at a marginal cost of capital, minus the present value of the cost of the investment. It can also be defined as NPV being the increase in the shareholder's wealth as a result of the investment at a given cost of capital. NPV is calculated as

$$NPV = \Sigma\ (PV_0 + PV_1 + PV_2 + PV_3 + PV_4 + \ldots + PV_n) \qquad\qquad (9.8)$$

where $PV_0 \ldots PV_n$ are the present values of cash flows in Years 0 to n.

Given that $PV_i = CF_i/(1+r)^i$

$$NPV = \frac{CF_1}{(1+r)^1} + \frac{CF_2}{(1+r)^2} + \frac{CF_3}{(1+r)^3} + \cdots + \frac{CF_n}{(1+r)^n} - I_0$$

$$NPV = \sum_{i=0}^{n} \frac{CF_i}{(1+r)^i} - I_0$$

where

I_0 – initial investment
r – cost of capital or the discount rate
n – project investment duration in years.

In the case study given earlier, the NPV calculation is given in Table 9.5.
For interpretation of NPV values the following rules of thumb are useful.

$NPV > 0$ Project financially *viable*. Financial returns greater than costs, value produced.

$NPV = 0$ Project financially *viable*. Financial returns equal to costs, value unchanged.

$NPV < 0$ Project financially *unviable*. Financial returns less than costs, value consumed.

In the example given the company's project increases shareholder value over and above the 20% cost of capital and therefore should be implemented.

It must be noted that NPVs for projects are more sensitive to changes in the discount rate than others, and this sensitivity will be dependent on the timing of the cash flows. Generally, projects with cash flows loaded towards the early years will be less sensitive to discount rates than if cash flows were loaded towards the later years.

Table 9.5 Net Present Value analysis of cereal company

Year	Cash flow	Discount factor (DF) at 20% discount rate $DF = 1/(1+0.2)^n$	PV Cash flows
0	−$1,300,000	1.0000	−$1,300,000
1	$420,000	0.8333	$350,000
2	$432,600	0.6944	$300,417
3	$445,578	0.5787	$257,858
4	$458,845	0.4823	$221,328
5	$472,714	0.4019	$189,973
$NPV = \Sigma PV_{(yr\ 1-5)}$			$19,575

9.5.5 Internal Rate of Return

Internal Rate of Return (IRR) is the discount or interest rate at which the net present value of an investment is equal to zero. In other words IRR answers the question 'What level of interest will this project be able to withstand?' Once this is determined the risk of changing interest rates conditions can be minimised. The IRR is the annual percenatage return achieved by a project, at which the sum of the discounted cash inflows over the life of the project is equal to the sum of the capital invested.

Microsoft Excel has a function that enables the calculation of the IRR. In the example given, the IRR is 20.7% which is greater than the organisation's cost of capital (20%). It can also be calculated manually by varying the discount rates in the NPV calculation until NPV = 0.

If the IRR is greater than the hurdle rate, the project needs to be accepted.

Whilst the NPV method gives an absolute dollar value, the IRR measure gives a relative measure of profitability.

The disadvantage of the IRR method is that it does not take into account the magnitude of the project. It only gives a relative measure of profitability as a percentage. The same IRR can be achieved by implementing a low NPV value project as well as a high NPV value project. For example, a project with a NPV of $100 and another with a NPV of $100 000 can have the same IRR. By implementing the second project it adds to the firm's bottom line. Therefore a choice based only on IRR alone may lead to erroneous ranking. Especially when assessing multiple projects it is useful to consider both the NPV and the IRR together to make an informed decision.

The objective of the organisation needs to be to select projects that adds to the bottomline not necessarily to maximise the profitability of individual projects.

9.6 Assessing Project Risk

In conjunction with assessing project financial feasibility, project risk too needs to be assessed. Risk management involves the identification, mitigation and evaluation of risks. As mentioned previously risks can be classified as

- financial
- technical
- regulatory (public safety)
- environmental.

The two features that characterise risks are

- the probability of an event occurring – likelihood
- the impact of the event on the project if the risk materialised – severity.

Table 9.6 Guidelines for the use of reclaimed water for industrial water systems

Type of reuse	Type of treatment	Reclaimed water quality	Reclaimed water monitoring	Controls
Closed system cooling system	Process specific	Site specific	Site specific depending on water-quality requirements and end use	Additional treatment by user to prevent scaling, corrosion and biological growth, fouling and foaming.
Open system cooling system (human contact possible)	Secondary and Pathogen reduction	Site specific Thermotolerant coliforms <1000 cfu/100 ml	pH, BOD and SS weekly Thermotolerant coliforms – weekly Disinfection – daily	Additional treatment by user to prevent scaling, corrosion and biological growth, fouling and foaming. Windblown spray minimised.

An accurate assessment of these two aspects will enable an organisation to understand the risks, evaluate the likelihood and severity scenarios, prioritise the risks and put in place actions to mitigate the highest risks.

Technical risk can be mitigated by selecting proven technologies, carrying out detailed laboratory and on-site pilot studies.

Financial risk can be mitigated by carrying out sensitivity analyses, hedging and having in place water-tight contracts with suppliers.

Regulatory risk can be mitigated by involving the relevant approving authorities early in the project-development phase. Water reuse projects may be subject to the state or national regulatory guidelines. These specify different requirements for water quality depending on the end use of water as well as the daily or weekly monitoring to be carried out by the project proponent [2–5]. Table 9.6 shows Agriculture and Resource Management Council of Australia and New Zealand *Guidelines for Sewerage System – Use of Reclaimed Water* requirements for recycled water projects for use in cooling towers and in the open systems where human contact is possible. Once the regulatory risks have been quantified then a risk mitigation methods needs to be developed and incorporated into risk management plan. The risk management plan should clearly spell out who is responsible to take corrective measures to address a specific event in the event of equipment failure.

References

[1] Gordon P.J. Executive Order, *Pollution Engineering*, pp. 20–22. March 2002.
[2] US EPA. *Water Reuse Manual*. September 2004.
[3] Agriculture and Resource Management Council of Australia and New Zealand. *Guidelines for Sewerage System – Use of Reclaimed Water*. November 2000.
[4] EPA Victoria. *Guidelines for Environmental Management – Use of Reclaimed Water*. March 2003.
[5] AWWA Research Foundation *Industrial Water. Quality Requirement for Reclaimed Water*. 2004.

Chapter 10

The Hospitality Sector

10.1 Introduction

Tourism plays a very important role in the world economy. The World Tourism Organisation estimates that there were more than 633 million international travellers in 1999 [1]. World international arrivals are expected to grow at a rate of 4.1% a year through the first two decades of the new millennium [1]. Asian tourism industry has the highest growth rates. The hospitality sector is one of the largest employers. For example according to the 2007 survey conducted by the National Restaurant Association there are over 12.8 million people employed in over 935 000 locations generating sales of US$ 537 billion.

This rapid growth in tourism will have a significant detrimental impact on the environment. In the United States alone, US$185 billion is spent on travel and of this US$37 billion is spent on accommodation [2]. More and more hotels will need to be built to cater for the increased tourist arrivals. On the other hand, operationally hotels are also large users of water per square area. Therefore, through water conservation, hoteliers can save money and the environment.

Some hotel resorts are located in remote areas where town water supply is non-existent and hotels depend on alternate supply sources such as bore water and tankers from nearby sources. Especially for these establishments security of supply is a critical issue and the financial benefits of water conservation are most compelling. In some of these locations such as in the Caribbean the cost of water can be as high as US$3.50/m³ (US$13/1000 US gal.) and if the water is delivered by tankers the cost of water can be as high as US$20/m³ (US$75/1000 US gal.).

Water conservation also provides other benefits, such as reduced infrastructure costs to store water, pumps, and reduced volumes of wastewater that need to be treated and discharged to the sewer or to the environment. In particular, remote locations can incur significant costs.

Some large hotel chains such as the Inter-Continental Hotel Group, Starwood, Hilton and Marriott participate in the UK-based International Hotel

Environmental Initiative (IHEI) to foster sustainable tourism practices within the tourism industry. Organisations such as the Green Hotels Association, the Coalition for Environmentally Responsible Economies (CERES) and the Green Hotels Initiative are spreading the benefits of *green hotels*. Hotel certification programmes such as Green Globe 21 provide hoteliers with marketable certifications. Not withstanding, water consumption per room still remains high in many hotels.

This chapter discusses water conservation opportunities in the hospitality sector which includes hotels, stand-alone restaurants and other food-preparation establishments.

10.2   Benchmarking Water and Energy Consumption

10.2.1   Benchmarking Water Consumption

Hotel water consumption can be benchmarked based on

- m^3/bed space/yr (Thames Water, UK)
- Water consumption per guest night (International Hotel Environment Initiative (IHEI))
- Water consumption per hotel per day (Sydney Water).

To gauge the factors that contributed to water consumption in hotels, UK water utility Thames Water carried out a survey in 1999 of 597 hotels located within its area of operations [3, 4]. The survey included information on the age of the building, catering facilities, the presence of swimming pools, leisure clubs, large gardens and air-conditioning systems. The study concluded that the main elements which affect water use in hotels are

- occupancy
- age of buildings
- hotel category (star rating)
- how efficiently water is used.

Occupancy has the largest bearing on hotel water consumption. A hotel category determines the type of facilities available, such as swimming pools and number of restaurants. Using the first two categories, Thames Water established benchmark figures based on water usage in *m³/bcd space/yr*. The study also concluded that ***three quarters of hotels*** used more water than the benchmark and about ***half of all the hotels in the area*** could save at least a quarter of their water bills by implementing some water-saving measures. Hotels with a very high consumption might save 50% or more. These results are shown in Figure 10.1.

The Thames Water benchmarks given in Table 10.1 provide a guide to water usage.

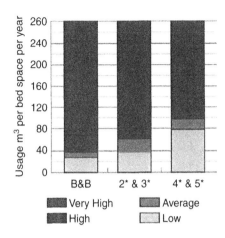

Figure 10.1   Water-use benchmarks for hotels

Courtesy of Thames Water, UK.

Table 10.1 Hotel water-usage benchmarks for water efficiency

Hotel category	Water usage per bed space per annum
Bed and breakfast	$30\,m^3$
2 and 3 star	$40\,m^3$
4 and 5 star	$80\,m^3$

Courtesy of Thames Water, UK.

The IHEI benchmarks are based on *litres per guest night*. The water usage is categorised under hotels less than 50 rooms, 50–150 rooms and greater than 150 rooms [5].

According to the US EPA the average hotel in the United States uses 791 L/room/day (209 US gal./room/ day). Studies done by the Seattle Public Utilities Board established that the median water usage in Seattle hotels is 545–719 L/room/day (144–190 US gal./day/room). Larger and more luxury hotels use 946 L/room/day (250 US gal./day/room) [6].

Sydney Water carried out 24 water audits in hotels in the Central Business District and developed benchmarks published in *Water Conservation Best Practice Guidelines for Hotels* [7]. The chart shown in Figure 10.2 gives best practice water usage for three different types of hotels based on occupancy and takes into account the presence of cooling towers and laundries (top line). Figure 10.3 shows the water-usage breakdown in each type of hotel.

10.2.2 Benchmarking Energy Consumption

The Australian Department of Industry Tourism and Resources carried out a survey of 50 hotels to gauge their energy consumption and develop

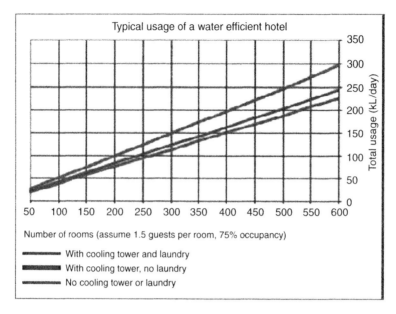

Figure 10.2   Charts to benchmark water usage

Courtesy of Sydney Water – Water Conservation Best Practice Guidelines for Hotels.

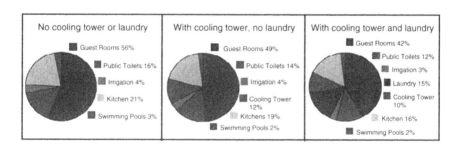

Figure 10.3   Typical water distribution in hotels [5]

Courtesy of Sydney Water – Water Conservation Best Practice Guidelines for Hotels.

benchmark indicators of best practice performance within the survey hotels [8]. The proposed benchmarks are as follows:

	MJ/Room (MM Btu/room)	MJ/m^2 (Btu/ft^2)
Accommodation hotels	35 000 (33 MM Btu/room)	750 MJ/m^2/yr (66 MBtu/ft^2/yr)
Business hotels	95 000 MJ/room (90 MM Btu/room)	1050 MJ/m^2/yr (92.4 MBtu/ft^2/yr).

Table 10.2 Typical Savings available per room

Area	Best Practice	Conventional	Savings per room per year	$ savings per year	Supply and installation cost	Payback period
Showers	9 L/min (2.4 US gpm)	22 L/min (6 US gpm)	36 m^3 (9511 US gal.)	A$117	A$30 per showerhead. Installation cost – A$30	6 months
Toilet	6/3 dual flush (1.6/0.8 US gpf)	11 L per flush (2.9 US gpf)	17 m^3 (4500 US gal.)	A$40.80	A$400–600 for new dual flush toilet	>10 years
Basin	6 L/min (1.6 US gpm)	12 L/min (3.2 US gpm)	5.3 m^3 (1400 US gal.)	A$12.70	Aerator A$5 Labour A$10 = A$15	Approximately One year

Water-efficient showerheads are available rated at 9–12 L/minute (2.4–3.2 US gpm) at normal pressures. In Australia under the WELS scheme only water-efficient showerheads are available from July 2006. Refer to Chapter 3 for more details on the WELS scheme and Table 10.2 for a calculation of water savings.

Some hotels are not keen to replace the showerheads under the misconception that there might be guest complaints or that it will damage their brand name. If the message is clearly communicated, guests will not complain and in many cases they are happy to stay in environmentally friendly hotels.

Sometimes the concern is that the showerhead quality must match with the hotel's image. In this case, flow restrictors can be installed without impacting on the aesthetics of the showerhead. A 300-room hotel on the Gold Coast in Australia installed 'Gem-flo' restrictors and was able to reduce the water usage from 18 L/minute to 12 L/minute (4.75–3.2 US gpm) without impacting on guest comfort. It costs the hotel $30 000 to retrofit but recouped this investment in 10 months. However in areas of inadequate water pressure, in-flow restrictors may not perform well. A better solution is well engineered showerheads.

Toilets. Poorly maintained toilets can leak a lot of water. Toilets with leaking flapper valves, overflowing tanks and defective flushing mechanisms are the frequent cause of leaking toilets. More importantly leakage from toilets can go undetected.

Older style cisterns use 12–14 L/flush (3.2–3.75 US gpf). If this is the case, these need to be retrofitted with more water-efficient models.

In 1984 the Australian manufacturer Caroma Industries pioneered the development of the *dual flush* toilet technology first by reducing the flush volumes to 9 L and then in 1993 launched the current industry standard

6/3 L/flush (0.8/1.6 US gpf) which saves water by offering a separate low flow setting for liquid wastes at 3 L/flush. The average flush volume is 3.8 L (1 US gpf). According to the manufacturer the potential water reduction from the traditional 11 L/flush to a 6/3 type is 67%. They have now followed up these developments with a 4.5/3 L/flush model (1.2/0.8 US gpf) which has the potential to reduce average flush volumes even further.

In the United States since the passing of the 1992 Energy Act for new toilets the minimum performance standard is 6 L/flush (1.6 US gpf) and are known as Ultra Low Flush (ULFT) toilets. However, the current industry push is to move to High Efficiency toilets (HET). To be sold as a HET, the toilet flush volume needs to be a maximum of 4.8 L (1.28 US gpf). Manufacturers and water utilities are also promoting a single full flush pressure–assisted toilets which has a flush volume of 3.8 L (1 US gpf). One reason for the preference for the single flush toilet over the dual flush model seems to be perceived user behaviour (which button to press). However, the pressure-assisted models are known to be noisier than the dual-flush variant.

In any event when retrofitting toilets using less than the standard, buyers need to satisfy themselves that the flow velocities with these ultra-efficient models are adequate to transport the wastes to the public sewer especially in instances with long drain lines and no additional fixtures nearby [10].

Items which will need to be maintained are the cistern rubber seals which need replacement every 2 years and a periodic inspection to replace valves and ballcocks in toilets. In flush-valve toilets inspect the diaphragms for correct operation.

Taps. Installing aerators or flow restrictors is a low cost but highly effective option. Aerators introduce air into the water stream to produce a larger and whiter stream. When installing flow restrictors be mindful that the hot water and cold water hydraulics are balanced. A disadvantage of flow restrictors are that since they have a fixed orifice, they produce higher flows at higher pressure (excess flow) and lower flows at lower pressure. One design that overcomes this limitation is the *pressure-compensating aerator* such as the Neoperl aerator featured in Figure 10.4. Pressure-compensating flow regulators maintain a constant flow regardless of variations in line pressure due to the specially engineered O-ring within the flow regulator that compresses or expands depending on the pressure. When the pressure increases the O-ring compresses against the seating area and relaxes when the pressure reduces as shown.

Leakage from taps can waste significant amounts of water. The following example shows the annual loss in a hotel with 150 rooms.

Leakage per minute (2 drips)	18 ml
Annual water loss	9.5 m³ (2 500 US gal.)
Percentage of rooms with leaking taps	25%
Total annual water loss from rooms	356 m³ (94 000 US gal.)
Cost (from Table 10.2)	A$850.00

Figure 10.4   Pressure-compensating aerators

Courtesy of Neoperl GMBH.

Assumptions

Water and wastewater charges	A$2.40/m^3
Occupancy	75%
Shower frequency	2/day
Duration of shower	5 minute/shower
Hot water costs included	
Labour rates	A$60/hr

10.3.3.2   Public Amenities
Public wash basins, toilets and urinals can account for 15–40% of water usage in a hotel. Options to increase water-use efficiency are detailed below.

Taps.   Install self-closing taps. The spring-loaded knob automatically shuts off the water when the user releases the knob. Ultrasonic-sensor taps allow water to flow for 10–15 seconds after operation. These are activated when the user's hands are placed beneath the tap; they shut off the water flow when the user's hands are removed from underneath the tap. Figure 10.5 shows a photo of an infrared tap. There is a debate whether these are superior to manually operated taps.

Urinals.   Of all the urinal types, the cyclic flushing units are the greatest water guzzlers. Single stall, manual flush cisterns are the most efficient. However, there is the risk of germs spreading from touching flush handles. If sensor-operated on-demand controls are preferred, then individual sensor flush units are a better option than a common sensor for a battery of urinals.

The Australian standard as defined by AS/NZS 6400:2005 is 1.0–2.5 L/stall (L/600 mm width of continuous wall). The US regulations for new urinals

Figure 10.5   Photo of an infrared tap

Courtesy of Zip Industries Pty Ltd.

require that the maximum water flow rate to be no more than 1 US gpf (3.785 L). The UK regulations are 7.5 L/hr.

Zip industries have recently launched the Zip Pearl Solo controller designed to service a single urinal. The system is easy to install. It is powered by a 6 volt lithium battery pack with a typical life of 3 years. An optional 240 volt mains power pack is available if required.

Operation is simple. Once a user has been standing within the range of the sensor for 5 seconds a passive infrared detector activates the flushing cycle. The fill-time for the cistern can be customized and the fast fill design helps eliminate scale problems associated with slow or drip-fed systems. An automatic janitorial flush every 12 hrs is also programmed into the unit.

A LED detector flashes to indicate that battery voltage is low or that the battery must be changed. In the event of battery power loss, the system automatically closes the latching valve to prevent water wastage. This feature is important, since with many sensor-operated systems one of the major causes of water leakage is sensor malfunction as a result of dead batteries. Figure 10.6 shows a photo of a Zip Pearl Solo sensor.

Caroma Industries, the developer of the 6/3 dual flush toilet, has recently launched Cube 0.8 L Smartflush®. It is ideal for institutions concerned about water conservation and yet achieve acceptable performance. The urinal has received the maximum Australian WELS 6 star rating. It uses only 0.8 L/flush without sacrificing performance. This is achieved by the urinal sensing the user and determines the flush mode accordingly. As a result it eliminates

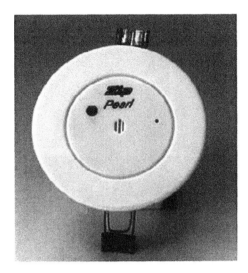

Figure 10.6   Zip Pearl Solo sensor

Courtesy of Zip Industries Pty Ltd.

unnecessary flushing. During peak periods it changes the mode to match the usage.

The water savings compared to existing stalls are as follows:

- water savings of 60% compared to standard 2 L single stall urinals
- water savings of 80% compared to a 2 stall urinal using more than 4 L/flush
- water savings of almost 90% compared to urinals using 7.5 L/flush.

These savings are seen from the chart in Figure 10.7.

Waterless urinals.   Waterless urinals have known to be in existence in Europe since 1885. Despite this only during the 1990s waterless urinals have come to the attention of water-conservation enthusiasts and cost-conscious users. To cite the water-saving potential of these units assume that a conventional unit flushes at the rate of 7.5 L/bowl/hr (UK regulations) then the water loss is 65 m^3/yr (17 350 US gal./yr). The annual cost of water and wastewater discharge at $ 2.40/m^3 equates to $157.

The benefits of waterless urinals are

- reduces water usage significantly.
- in cold climates require no freeze protection.
- eliminates overflows from urinal bowls when blocked.
- reduces costs from usage and water leaks.
- reduces the need to upgrade booster pumps in high rise buildings.
- essentially maintenance free apart from the regular cleaning.

Caroma continues to lower urinal water consumption

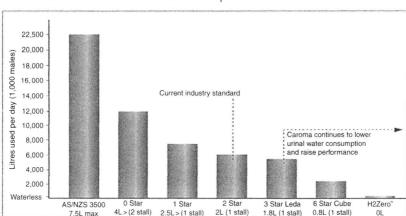

Figure 10.7   Water saving with Cube 0.8L Smartflush®.

Courtesy of Caroma Industries Ltd.

- eliminates pipe blockage in hard water areas due to calcium reacting with urine scale.
- no flush controllers to maintain and batteries to replace in sensor units.
- reduced wastewater treatment plant loads.
- enhances the green image of the organization.

There are a number of waterless urinal technologies in the market. Some include

i)   vegetable- or alcohol-based odour barrier in a *disposable cartridge* or *fixed trap*
ii)   microbial blocks – bacteria or enzyme seeded
iii)   mechanical working traps.

Waterless urinals make use of the fact that urine is a liquid and therefore no additional water is required to flush it down the drain.

The most common is the cartridge type. These units are made of acrylic, fiberglass or ceramic. The cartridge is inserted in the urinal and connected to the drain pipe. When urine flows into the cartridge, it passes through a liquid sealant that is less dense than the urine but is more viscous. Therefore the sealant floats above the urine forming an odour barrier between the urine and the air, preventing the odours from escaping into the air. The urine then flows into a central chamber and into the sewage line. Sealants are vegetable oils or alcohols, both types have a density less than water. This type of waterless urinal is marketed by suppliers such as Uridan and Ernst. A typical bank of waterless urinal units and a waterless trap is shown in Figures 10.8 and 10.9. A problem with these units is that if excess water is poured it runs

Figure 10.8   A bank of Uridan (fixed cartridge type) waterless urinals

Courtesy of Watersave Australia Pty Ltd.

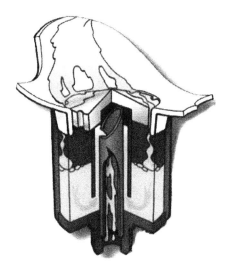

Figure 10.9   Waterless trap

Courtesy of Watersave Pty Ltd.

the risk of losing the oil sealant. This has the potential to cause odours since the sewer gases are no longer prevented from contaminating the building environment.

Disposable cartridges type – Manufacturers recommend that the disposable cartridges be replaced after approximately 6000–7000 uses. In practice these are replaced more often (as low as 2000 uses) which impacts on costs. Used cartridges are a health and safety hazard and therefore they need to be disposed correctly.

In the fixed trap system, only the liquid sealant needs to be topped up after approximately 5000–7000 uses by pouring a measured amount of liquid to the cartridge.

The cartridge has the added benefit of also preventing foreign objects such as cigarette butts being put down the drain.

Caroma Industries have released a new type of trap called the H2Zero™ Cube. It operates utilising unique patented cartridge technology that does not use an oil-based seal as used in traditional waterless urinals. Housed within the cartridge the Bio Fresh deodorising block is activated during use, while the Bio Seal™ allows urine to pass through the seal freely. The Bio Seal™ acts as a one-way air-tight valve to seal the cartridge from the drainage system and against back-pressure situations. The manufacturer claims that this operation guarantees consistently superior performance and hygiene compared to other waterless urinals. Figure 10.10 shows this unit.

Bacterial blocks.   The bacterial blocks are made from naturally occurring bacteria in a urine-soluble block. These blocks are placed in conventional urinals (the water supply is shut off) and upon contact with urine dissolves releasing the bacteria. The manufacturers claim that the bacteria decompose the urine to a non-odorous non-scaling form. It also releases a citrus smell. Maintenance is confined to spraying the urinal with a bacterial booster to increase the breakdown rate of bacteria as well as pouring a little water down the urinal on a daily basis to minimise uric acid formation. Their biggest advantage over the cartridge type is that they do not require retrofitting of the existing urinal.

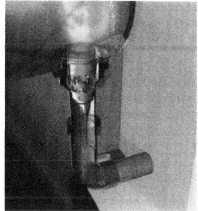

Figure 10.10   H2Zero™ Cube

Courtesy of Caroma Industries Ltd.

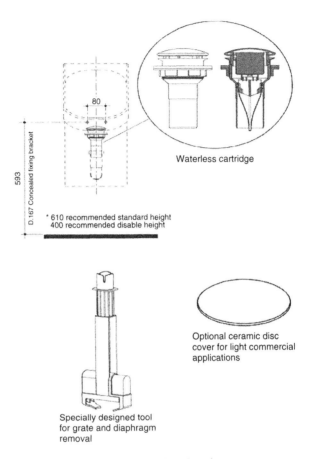

Waterless cartridge

593
D.167 Concealed fixing bracket

80

* 610 recommended standard height
400 recommended disable height

Optional ceramic disc
cover for light commercial
applications

Specially designed tool
for grate and diaphragm
removal

Figure 10.10 (Continued)

Ongoing costs are for the purchase of bacterial blocks (which can last up to 2 weeks) and bacterial booster. In some cases the costs have been greater than the water savings.

Mechanical working traps – trap with duck bill.   Suppliers such as Sphinx offer a waterless urinal known as duck bill where a hose made from special waterproof material serves as a trap. Its opening is round at the top and pressed together. It lets the urine through and immediately closes shut afterwards. This type of trap come without the floating sealing liquid nor need power to activate the device. Since there are no moving parts it minimises the potential for maloperation or trap failure.

User acceptance.   The acceptance of these technologies has been mixed. Early concerns of urine smell are no longer an issue with these units if regular cleaning in accordance with the manufacturers' recommendations is carried out, that is by spraying the bowl with a disinfectant daily and wiping it down with a soft cloth. In hard-water areas, to prevent scale build up it is suggested cleaning with two liters of warm water mixed with detergent every week. There is some concern that without a permanent water seal in the urinal sewer gases could potentially pose threats to public health from toxic gases including hydrogen sulphide, methane and airborne pathogens. Consequently the International Association of Plumbing and Mechanical Officials (APMO) Standards Council have held off approving Waterless Urinals.

Researchers have also cast doubts on whether 2 L per week in the absence of any other fixtures being connected to the sewer line is sufficient to prevent struvite formation in them.

The following factors need to be considered for successful installation of waterless urinals:

- Ensure correct drain pipe material and slope – Drain lines must slope at least 2 cm/m (1/4 in./ft). The drain lines require adequate venting for trouble-free operation. For these reasons, older urinals built for water flushing may not be always suitable for retrofitting with waterless urinals.
- Ensure that other water fixtures are present to minimise sedimentation – Professor Mete Demiriz of Gelenkirchen University, Germany investigated a number of liquid seal and cartridge-type urinals over a 2-year period and concluded that there is sedimentation in the drain lines which cannot be removed by simply adding high water volumes through the urinal [11].
- Replace copper piping – If the drain lines are of copper these need to be replaced with non-metallic sanitary plumbing to prevent the breakdown products of urine (ammonia) corroding copper piping in accordance with plumbing standards such as the Australian AS/NZS 3500.
- Eliminate drain pipe obstructions – If retrofitting to existing systems, best to clean the sewage line with a power cutter before installation.
- Maintain a urinal maintenance log – Maintain a log book to record dates of cartridge purchases and their replacement.

Costs and savings.   Initial installation and annual operating costs can vary by vendor. The cost savings achievable from these units are dependent on

- The type of waterless urinal installed. For example, the microbial blocks have minimal installation costs but the ongoing costs can be more than for the fixed cartridge type. Whereas, with the cartridge type, retrofitting will incur additional costs.
- The flush volume of the urinals that are replaced.
- The number of users per day.
- The pricing structure for water and sewer usage.
- The maintenance costs (dependent on the cost of cleaning and cartridge seal replacement).

In general, it is accepted that the cartridge type will provide a payback within 3 years when retrofitted into an existing urinal.

10.3.3.3 Kitchens

In many instances, the water-saving potential in kitchens are ignored. Kitchens account for 7–20% of water usage in a hotel. Savings in water usage also realises energy savings.

Within the hospitality sector, restaurants have the highest energy intensity. US data suggests that restaurants on average use 2.84 GJ/m² (250 MBtu/ft²), roughly 2.5 times more energy per square area than commercial buildings [12]. Food preparation, refrigeration and sanitation represent nearly 60% of the energy consumed in a typical food service facility. The total energy savings potential in a commercial kitchen can vary from 10 to 30%, depending upon the technologies installed.

Benchmarking water usage using industry standards is a first step towards understanding water-saving potential in a commercial kitchen.

The industry standard is *liters per food cover*. A *'food cover'* is defined as any transaction or sale, whether a cup of coffee or a multiple course meal. Sydney Water recommended L/food cover for hotel situations are [7] given in Table 10.3.

These figures will need to be multiplied by a factor to account for the different types of restaurants such as Chinese and stand-alone A-la carte restaurants. Inefficient uses of water in kitchen operations arise mainly from two areas: equipment **design and behavioral patterns**. The water-saving opportunities in kitchens are

- sinks
- dishwashers
- Asian wok cookers
- ice makers
- garbage disposal units
- leakage.

Sinks.   Below are some easy and practical ideas to save water and energy to maximise your profits.

- Turn off taps when not in use.
- Minimise leakage by replacing worn washers.

Table 10.3 **Water efficiency per food cover**

Rating	Water Usage (L/food cover)
Good	< 35
Fair	35–45
Poor	> 45

- Water usage in sinks and basins can be halved by installing aerators or flow control regulators. A tap using 25 Lpm (6.6 US gpm) as against a tap rated at half the amount can waste 40 m³/yr (10.5 thousand US gal./yr).
- Install infrared motion-activated sensors or pedal-operated taps.
- Do not thaw frozen foods or wash rice under running water. If thawing is done under running water, assuming a minimum of 1 hr/day at a flow rate of 18 Lpm up to 591 m³ (156 000 US gal.) per year could be wasted. At $2.40/m³ this could amount to $1419/yr. Use a stand-alone thawing unit or the bottom shelf of refrigerator.
- When washing pots fill sinks for washing pots instead of running water.
- Educate your staff on saving water and energy.

Low-flow pre-rinse spray valves.   Pre-rinse spray valves (PRSV) are used to pre-rinse dishes and pans before loading to a dishwasher. In western style commercial kitchens these units can account for a significant amount of water usage. PRSVs consist of a spray nozzle, a squeeze lever that controls the water flow, and a dish guard bumper. Models may include a spray handle clip, allowing the user to lock the lever in the full spray position for continual use. In conventional PRSVs, water flows at low velocity in a circular pattern from multiple holes similar to conventional showerheads. These can use 10–20 Lpm (2.64–5.2 US gpm) of water with an average flow rate of 13.7 Lpm (3.6 US gpm) [13].

In North America [13–16] due to the efforts of Food Service Technology Center (FSTC), and the California Urban Water Conservation Council, low flow pre-rinse spray valves have become a viable water conservation measure in commercial kitchens. Unlike the conventional PRSVs, low-flow PRSVs produce a fan like spray using just 6–8.3 Lpm (1.6-2.6 US gpm) with an average flow of 5.6 Lpm (1.4 US gpm)[13]. Figure 10.11 shows photos the difference in flows between the water efficient model from those of the conventional models.

A *'cleanability test'* has been developed to qualify as a low flow model. The American Society of Testing and Materials *ASTM F2324–03* requires that the spray valves be able to clean plates in 21–30 seconds per plate at a pressure of 406 kPa (60 psi). In Australia, low flow PRSVs have been added to the mandatory WELS scheme. In addition to saving water, these units also reduce gas or electricity. The quantum of water savings correlates with hours of usage. Very small establishments could use for less than an hour to large users using greater that 4 hours per day. As Table 10.4 shows the water and energy savings achievable when the units are used for 2 hours per day is A$5020 with a payback of 15 days.

It is worth remembering that high water pressures in excess of 551 kPa (80 psi) could potentially cause splashing and low water pressures less than 276 kPa (40 psi) will result in poor cleanability and dissatisfaction with the unit. Therefore check the line pressure before installing a low flow variant.

Figure 10.11   A photo of a low flow pre-rinse spray value in action

Courtesy of Niagara Conservation Inc.

Table 10.4 Water and Energy Savings from low flow pre-rinse spray valves

Performance	Base model	Best available
Nominal flow rate at a pressure of 406 kPa (60 psi)	13.7 Lpm (3.6 US gpm)	6 Lpm (1.6 US gpm)
Annual water use (usage is 2 hrs per day for 313 days of the year).	514 m³ (135,950 US gal.)	227 m³ (60,096 US gal.)
Annual water and sewer costs at $2.40/m³	A$2,364	A$1,282
Annual energy use.	94 GJ	42 GJ
Increase in water temperature 31° C.		
Heater efficiency at 70%		
Annual gas cost at A6/GJ	A$564	A$249
Water and energy costs based on a 5-year life of equipment.	A$8,996	A$3,977
Water and Energy cost savings over the 5 years.	–	A$5,020
Installed cost	A$214	A$214

Asian wok stoves.   Asian wok stoves are typically the heart of any Chinese restaurant's kitchen. Studies conducted by Sydney Water's Every Drop Counts Business Program have shown that these stoves waste as much as 5000 L/day (1,321 US gal./day) [17]. By installing a *waterless wok stove* the initial capital can by recouped in 1 year. This is explained below.

Conventional wok stoves use water for two main purposes:

1.   As cooling water to cool the cook top of wok stoves to prevent the metal from overheating and warping. Therefore water flows at three to four liters per minute, which adds up to 2500–3500 L/day going to drain.

2. Rinsing of the wok between dish preparations requires another 2000–3000 L per day. These swivel taps are not shut off when not in use.

Recognising these two problems, Sydney Water pioneered the development of a *waterless wok* stove that is air cooled with special swivel tap that shuts off when pushed to a side. The resultant cost benefits are

- Water savings of 5000 L/day/stove. Annual water savings of 1800 m^3/yr.
- Monetary savings of $4500/yr based on water and sewer usage costs of $2.40/m^3 and 360 days of operation per year.
- Kitchen grease traps are also not overloaded since less water goes to the drain. However, since the grease traps are working more efficiently the grease trap will need to be emptied more frequently.

Top photo of Figure 10.12 shows the constant water flow required to cool a conventional wok and the spout also wasting water. The schematic shows the air flow pattern in these units. Figure 10.13 shows the absence of water on the hot metal plate in a waterless wok, the spout doesn't waste water and the schematic shows how the air flow keeps the units cool. The graph in Figure 10.14 shows the reduction in costs per food cover for the two types of woks.

Dish washers. Commercial kitchens have multiple dishwashers. These dishwashers can be operating from 1 hr to 16 hrs a day. In such instances, they can waste a lot of water if not maintained correctly.

Essentially there are four types of dish-washing machines in commercial kitchens. These are classified as "undercounter, door, conveyor and flight". Irrespective of the type they all have wash, rinse and sanitising cycles. The water usage and capacities of these four types are given in Table 10.5

The following are the water- and energy-saving techniques for dishwashers.

- Train staff employees on the importance of operating the dishwasher at full load.
- Ensure that water jets are not missing. The water jets need to be replaced periodically. The water flow quadruples with a doubling of the diameter of the nozzle.
- Use low-flow pre-rinse sprays to remove food scraps before loading a dishwasher.
- Use a water-efficient machine if possible.
- Recycle final rinse water if equipped to do so.
- Stick to the manufacturer's recommended flow rates.
- Install electric *eye sensors* to run the machine only when dishes are present. The sensor feature helps maintain the temperature in the tanks and extends machine life by reducing the pump operating time.

Conventional Wok Stove

Water-cooled wok stove

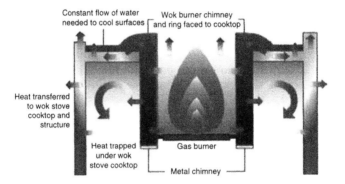

Figure 10.12   Conventional wok stove and air-flow pattern

Courtesy of Sydney Water.

- Reuse the rinse water from the dishwasher as flush water in garbage disposal units.
- Use water from the steam table, instead of fresh water, to wash down the cook's area.
- Investigate if the local water or energy utility gives rebates for replacing older (water and energy) inefficient models.

Here are some points to remember about chemical sanitisers. Using low temperature or chemical sanitizer machines is one way to save on hot water costs and reduce some ventilation requirements. Chemical sanitising machines use a sanitising chemical in the final rinse rather than hot water to do the job. While chemical sanitisers reduce energy costs, this needs to be weighed up with increased chemical costs and 'cleanability'. High-temperature machines are better able to break down animal fats and grease as well as lipstick on glassware and dishes. If the active ingredient in the

Waterless Wok Stove

Air-cooled wok stove

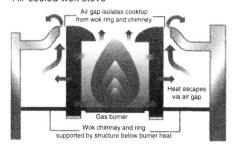

Figure 10.13 *Waterless wok* stove and air-flow pattern

Courtesy of Sydney Water.

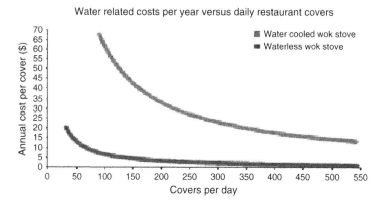

Figure 10.14 Water-related costs per year of a traditional and waterless wok stove

Courtesy of Sydney Water.

sanitiser is bleach (sodium hypochlorite), this may attack certain materials, including silver, aluminium and pewter.

To adapt an old cliché, *take care of your dishwasher and it will take care of you.*

Table 10.5 Dishwasher types and water usage

Type	Description	Typical Fresh Water Usage per rack
Undercounter	40–50 settings per hour. Suitable for bars and snack bars etc. Highest water user.	2.5–10 L (1.1–2.6 US gallons)
Door	90–110 settings per hour. Most widely available. Seen in small-to medium-size restaurants. Some models are equipped with recycling of rinse water.	3.5–9.5 L (0.9–2.5 US gallons)
C – Line Conveyor Rack	> 200 settings per hour. Most widely used in hotels. Equipped with separate pre-wash, wash and rinse compartments. A pre-wash helps minimise detergent use by not introducing it until heavy soils are removed from the china. The next higher capacity and better performance is the addition of a rinse tank or an extended wash tank. A rinse tank gives the ability to recirculate water and therefore conserves water. All machines have a fresh water final sanitising rinse.	3.0–6.0 L (0.8–1.6 US gallons)
Flight	High volume washing capability 14 000 dishes per hour. Dishes loaded directly on to belts. Comes equipped with electric eye sensors so that machine only operates when the belts have dishes.	2.5–9.5 L (0.7–2.5 US gallons)

10.3.3.4 Cooling Tower: Air-Conditioning and Refrigeration

The hospitality sector is dependent on air-conditioning and refrigeration for guest comfort and for maintaining food at the right temperatures. They also account for 47% of energy use in a typical hotel. Generally, hotels with extensive air-conditioning use 50% more electricity than those without, and also use more for heating, given the extra ventilation needs [18]. Cooling towers are a necessary component of a large air-conditioning system and was covered at length in Chapters 5 and 6.

Here are some specific tips to reduce the air-conditioning and refrigeration loads in the hospitality sector. Reducing the air-conditioning and refrigeration loads reduces cooling water requirements when they are connected to water-cooled condensers.

- When setting up a function, make certain that the cooling and lighting are off until 1/2 hr before the start of the function.
- Keep refrigeration doors closed. Use strip curtains or plastic swing doors on walk-in refrigerators and freezers. These measures will reduce compressor run time.
- Replace worn door gaskets in refrigerators.

10.3.3.5 Laundry

Some hotels have on-site laundries. Typically they are of the washer extractor type rather than tunnel washers. Water conservation opportunities for laundries are covered in Chapter 15.

10.3.3.6 Ice-Making Machines

Ice-making machines are common in restaurants, hospitals and hotels. They are noted for their high consumption of electricity.

Ice-making machines use water in two ways: for cooling the machine and for freezing water into ice. It is estimated that a water-cooled ice machine producing 363 kg (800 lbs) of ice per day and running at 75% capacity will consume about 3400 L/day (900 US gal./day) just for cooling [19].

There are four categories of ice makers:

1. Cube – clear, regularly shaped ice of a certain weight. The vast majority fall into this category.
2. Flake – ice formed into chips or flakes that contain up to 20% liquid water.
3. Crushed – ice that consists of small, irregular pieces made by crushing larger chunks of ice. Primarily used to cool drinks.
4. Nuggets – ice made by extruding and freezing slushy flake ice into small pieces.

Ice *cube* makers use more water than *flake* ice machines because a large proportion of water used in ice cube makers is used to purify the ice by removing the minerals in water to produce clear ice cubes. Flaky ice is not required to be crystal clear and therefore the minerals that give rise to cloudiness need not be removed.

A typical ice-making machine has the following components:

- a condensing unit used for cooling
- an evaporator surface for ice formation
- an ice harvester
- an ice storage container.

The condensing unit can be air-cooled or water-cooled as described below.

There are three types of ice makers:

1. ice-making head
2. self-contained
3. remote condensing unit.

Water usage can vary from 110 to 2500 L to produce 100 kg of ice.

Types of condenser.   Ice makers are available with two types of condensers as discussed below.

Air-cooled condensers.   They are the most energy intensive of the models but use the lowest amount of water. Energy usage varies from 44 to 180 MJ to produce 100 kg (5.4–22.5 kWh/100 lb) of ice. Air-cooled condensers can be found in ice-making head, self-contained units or remote condensing units. If the air-cooled condensers are located inside the building it adds to the air-conditioning load of the building. Remote condensing units are split-system models in which the ice-making mechanism, the condensing unit and the ice storage bins are in separate sections. Remote air-cooled condensers transfer heat generated by the ice-making process outside of the building. Like water-cooled units, they reject heat outside of conditioned spaces and therefore do not increase air-conditioning loads. They also reduce noise levels inside by up to 75%, but there are extra installation costs for running lines to a remote location.

Water-cooled condenser models.   They are more energy efficient than air-cooled units, using 37–113 MJ/100 kg (4.7–14.2 kWh/100 lb) of ice. There is no addition to air-conditioning loads, because the heat removed in making the ice is discharged outside the building. However, as shown in Figure 10.15, water usage is nearly 10 times that of air-cooled machines [19, 20].

Self-cleaning models use significantly more water than others. Typically, ice makers are cleaned and sanitised every 2–6 weeks, which requires emptying the bin of ice, adding cleaning solution, switching the controls to a cleaning mode that circulates the cleaning solution through the machine and then producing enough ice to be sure the machine is cleared of the solution. While self-cleaning models automate most of these steps they consume more water.

Water conservation tips for ice machines are as follows:

- Water-cooled ice makers need to be connected to a cooling tower so that the condenser cooling water does not go to the drain. Single-pass cooling needs to be avoided.
- Use the minimum flow rates for condenser cooling as per the manufacturer's guidelines.
- If cooling water cannot be connected to a cooling tower, then consider the possibility of reusing the cooling water in a non-potable application.
- Select ice-flaking machines over ice-cube machines since these require less water. The US Department of Energy's Federal Energy Management Program has produced a fact sheet *How to Buy an Energy Efficient Commercial Ice Cube Machine*. This can be downloaded from www.eren.doe.gov/femp/procurement.

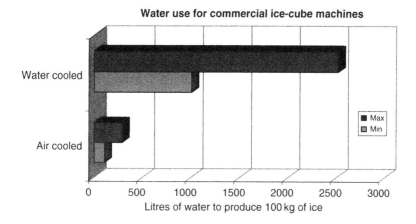

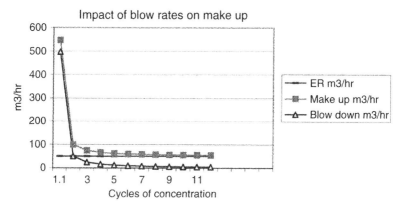

Figure 10.15   Water usage in commercial ice-making machines

Adapted from North Carolina Department of Environmental and Natural Resource *Water Efficiency Manual for Commercial, Industrial and Institutional Facilities.*

- Typical useful life of an ice-making machine is 5 years. If replacing older machines do not oversize machines, since this increases purchasing costs as well as energy and water costs and usage. Consider the possibility of air-cooled condensers over water-cooled condensers. However, be aware that air-cooled condensers produce less ice than water-cooled models as well as consume more electricity.
- For ice-cube makers use softened water. Softening the water removes the minerals that leads to cloudiness and therefore less bleed-off is required from the machine.
- Train employees to use ice only when required.

10.3.3.7   Swimming Pools

It is difficult to find a hotel, motel or club without a swimming pool of some sought, in the scheme of things these do not use large amounts of water unless there are leaks. From the vast number of water audits of hotels and clubs carried out by Sydney Water, the percentage of water in use in swimming pools is only around 3%.

Recommended actions will be covered in Chapter 12.

10.3.3.8   Staff Rooms

Retrofit shower heads, taps and basin taps with water-efficient models.

10.3.3.9   Irrigation

Many resort hotels have large gardens, golf courses that consume significant quantities of water.

There are many good publications on water efficiency for landscapes such as *Handbook of Water Use and Conservation* by Amy Vickers. The suggestions given below are not exhaustive. Water utility web sites too have information specific to the location and climate. The reader is directed to refer to these sites for more details.

However some principles are given below:

When discussing water efficient landscapes it is important to recognize the trifecta of:

- Right plants
- Right irrigation
- Right soil

An over reliance of one factor over the other two does not produce the optimum results.

Right plants

1. Plant selection to suit the climate, site layout (includes slope, degree of shading) and existence of other plants. Consider the degree of traffic.
2. Group plants with similar water needs together. Some plants need more water than others. High water demanding plants to be planted at the bottom of slopes.

   Choose plants with adaptations that make them natural water savers. Give preference to indigenous plants. In general, plants with hairy, succulent, wax-coated leaves or with fine, stiff foliage (sclero-phyllus) are adapted to growing in dry environments. Water saving grass such as Nioka and Palmetto have deep roots and are therefore drought tolerant. They also don't need mowing as frequently as other lawns.

Grey-or silver-foliaged plants are also usually suited to dry conditions. Sydney Water has an online plant selector to help consumers select the right plant combinations for the location.

Right irrigation

Right irrigation can be described as the right amount of water at the right time. Efficient irrigation practices consist of:

1. Right watering practice – water when evaporation is lowest – early morning and evening. About 10 mm (sandy soils) to 30 mm (loamy soils) of water should be applied to wet the soil to a suitable depth.
2. Right equipment – for plants - drip irrigation systems that water the root zone. Low volume pop-up sprinkler heads for lawns. Use the correct pressure and select the right nozzle size and diameter of pipe.
3. Right spacing between sprinklers - In general, the spacing between sprinklers should be about 50 to 60 percent of the wetted diameter. Minimise overspray of concrete and other paved areas.
4. Right scheduling of irrigation - Automatic systems and rain and moisture sensors to increase the efficacy of the irrigation system. A rain sensor attached to a controller catches moisture and prevents the sprinkler system to water in the rain. Timers irrigate according to a preset hours. SCADA control systems can be used for large lawns such as in parks, universities etc. Soil moisture sensors overrides the system when soil moisture is adequate. These are more accurate than rain moisture sensors since they measure the moisture in the root zone. All of these sensors are useful if the system can be monitored and adjusted regularly.

Weather based evapotranspiration controllers (ET) are devices that estimate the watering needs of a particular site based on the formula given in (10.1).

Water requirement = (potential – actual) evapotranspiration.(10.1)

These range from stand alone controllers based on local site sensors to controllers with remote programming ability.

Right soil

It is important to ensure that the soil is suitable for the plants. The chemical and physical attributes such as pH, permeability and water retention capabilities are just as important as the right irrigation equipment. Mulching is a proven method to reduce the natural evaporation of water. Organic mulches add nutrients and humus to the soil as they decompose, improving it and its moisture-holding capacity and reducing the need for watering and maintenance. Mulches range from grass clippings to Lucerne and sugar cane. The last two protect against nitrogen draw off from the soil.

10.4   Staff Awareness Programmes

Staff awareness programmes play a crucial role in a holistic water conservation programme. Behavioural change programmes among employees lead to water savings at minimal cost; however, it is difficult to predict their savings potential beforehand. The Westin Hotel in Seattle, United States, had achieved 6% reduction in water savings amounting to 38 kL/day (10 000 gal./day) through behavioural measures [21]. However, when designing a behavioural change programme, the educational and social backgrounds of the employees may require innovative approaches. For example, in the Australian catering and restaurant industry, 70% of the workforce have no post-school qualifications and 48% of the workforce all work either part-time or casual [22]. If the majority of the employees are from non-English-speaking backgrounds, the information may need to be translated into suitable languages.

A summary of potential behavior measures are listed below:

- *Reduce or discontinue triple sheeting* – No cost. Leads to water, chemical and energy savings.
- *Train kitchen staff* to correctly thaw frozen food, rice rinsing, cleaning of dishes using pre-rinse spray valves, dishwasher loading and equipment cleaning. Minimal costs with immediate paybacks.
- *Train housekeeping staff – to reduce water usage during cleaning of rooms* – It is common practice for cleaning staff to flush the toilet as they walk into a guest room irrespective of the cleanliness of the toilet bowl. After cleaning, the toilet is flushed again: By flushing the toilet only after cleaning (providing the bowl is clean to start with), the Westin Seattle saved on average 11.3 kL/day (3000 US gpd) [21].
- *Provide ice water only on request* – Restaurant staff bring ice water and refill guest glasses with ice water irrespective of whether the guest requested it or not. By stopping this practice, water and energy can be saved.
- *Wash floors, loading docks and rolling carts* – Train staff to use water judiciously. Use a broom where possible.
- Display the current water and energy consumption per room or guest night charts so that they can see the fruits of their labour.
- *Reward employees for doing the right thing* – Through newsletters, bulletin boards, websites and meetings. Publicise their efforts. Give a cash bonus incentive if the department's target water usage is achieved. Send employees on appropriate training courses.

10.5   Guest Awareness Programmes

Towel-linen exchange programmes are a well-accepted guest awareness programme, where a guest staying for more than one night is encouraged to reuse the towel and bed sheets rather than being changed daily. Even with only a 25% participation The Sheraton Rancho Cordova has seen a 5% reduction in water usage [21]. These programmes are more prevalent in hotels where the majority of guests are for business. However, their wider use is hindered by a number of preconceived notions. These include

- fear that the guest will complain
- guests may perceive the hotel as being *cheap*
- previous guest complaints
- corporate policy against towel-linen programmes.
- belief that daily linen service is commensurate with room price.

Rather than being considered *cheap* guests, support such initiatives. A Holiday Inn study showed that 80% of their business and leisure travellers were more inclined to stay at a place where a towel-linen programme was in place. Increasingly travellers are adopting *value based behaviour* to rate hotels. Thus they can align personal values with business values by giving their business to those that are striving to make the planet a better place to live.

Case Study

The Fairmont hotel chain is the largest luxury hotel operator in North America. In recognition of their commitment to the environment the Fairmont Hotels & Resorts (BC Region) was awarded the *2005 Energy & Environment Award* from the Hotel Association of Canada. The press release states,

"This award is given to a lodging property that has developed a culture towards integrating environmental management practices that improve everyday operations and the bottom line, while maintaining quality service and meeting guest expectations".

Among other environmental initiatives they have a policy of informing the guest that: "To reduce the negative impact on our environment through the use of laundry detergent chemicals and energy consumption, it is our practice to change bed linen once every three days or upon request."

Adapted from: *www.waterinthecityvictoria.ca.*

References

[1] The Ecotourism Society. *Ecotourism Statistical Fact Sheet.* 2000.
[2] Coalition for Environmentally Responsible Economies. Green Hotel Initiative. www.ceres.org/our_work/ghi.htm. October 10 2002.
[3] Thames Water. *Water Efficiency in Hotels.* www.thameswater.co.uk.

[4]　Waggett R. and Arotsky C. *Water Key performance indicators and benchmarks for offices and hotels*. CIRIA. London. 2006.

[5]　International Hotels Environmental Initiative. *Environmental Management for Hotels – The industry guide to best practice*. Butterworth-Heinemann. Second edition 1996.

[6]　Seattle Public. Utilities *Hotel Water Conservation – A Seattle Demonstration*. p. 4. July 2002.

[7]　Sydney Water. *Water Conservation Best Practice Guidelines for Hotels*. Sydney. December 2001.

[8]　Department of Industry, Tourism and Resources. *Energy Efficiency Opportunities in the Hotel Industry sector*. www.industry.gov.au/assets/documents/itrinternet/Hotels Benchmarking Report20040206161738.pdf. April 2002.

[9]　Energy Efficiency Best Practice Programme, Energy Efficiency in Hotels – A Guide for Owners and Managers. Department of the Environment. Walford, UK. 1997.

[10]　Koeller J. *High Efficiency Toilets (HET™)*. www.awwa.org.

[11]　Demiriz M. *Application of Dry Urinals*. Gelsenkirchen University of Applied Sciences. 45877 Gelsenkirchen, Germany.

[12]　Consortium for Energy Efficiency Inc. *Fact sheet – Commercial kitchens*. www.cee1.org.

[13]　Veritec Consulting Inc. City of Calgary Pre-Rinse Spray Valve Pilot Study. Mississauga, Ontario. December 2005.

[14]　Department of Energy, Federal Energy Management Program. *How to Buy a Low-Flow Pre-Rinse Spray Valve*. WS – 5. September 2004.

[15]　SBW Consulting Inc. Evaluation, *Measurement & Verification Report for the CUWCC Pre-Rinse Spray Head Distribution Program*. Bellevue, WA 98004, May 3 2004.

[16]　Bing T. and Koeller J. *Pre Rinse Spray Valve Programs: How Are They Really Doing?* SBW Consulting Inc. December 2005.

[17]　Sydney Water. *The Waterless Wok Stove*. www.sydneywater.com.au.

[18]　Department of the Environment. *Guide 36 – Energy Efficiency in Hotels – A Guide for Owners and Managers*. The Carbon Trust. London, UK. 1997.

[19]　New Mexico Office of the State Engineer. *A Water Conservation Guide for Commercial, Institutional and Industrial Users*. July 1999.

[20]　North Carolina Department of Environmental Natural Resources. *Water Efficiency Manual for Commercial, Industrial and Institutional Facilities*. August 1998.

[21]　Seattle Public Utilities. *Hotel Water Conservation – A Seattle Demonstration*. July 2002.

[22]　Restaurant & Catering Association. *Industry Facts*. www. restaurant-cater.asn.au.

Chapter 11

Commercial Buildings, Hospitals and Institutional Buildings

11.1 Introduction

Commercial and institutional buildings – such as office buildings, shopping malls, universities, hospitals and prisons – can account for significant water consumption. In California, for instance, 27% of the State's urban water supply (of 7 million acre – feet (8633 GL/yr)) is used by the commercial and institutional sectors, which is nearly three times that of the industrial sector [1]. The World Business Council for Sustainable Development states that buildings are one of the largest end users of energy; in OECD countries, the building sector accounts for 25–40% of the final energy demand and this demand is expected to grow by 45% by 2025 [2]. Out of this, 33% is used by commercial buildings.

Therefore the commercial and institutional property sector needs to respond to these needs by becoming energy- and water-efficient. Most of the fixtures in commercial and institutional buildings are similar (with slight variations) and are therefore considered together in this chapter. The solutions for water savings are similar for hotels (in most cases) and (where applicable) the reader is referred to the relevant sections in Chapter 10.

11.2 Commercial Property – Office and Retail

11.2.1 Industry Structure and Water Usage

The commercial property sector consists of offices, shopping centres (retail), industrial parks, hotels and car parks. Trusts known as Listed Property Trusts (LPT) or Real Estate Investment Trusts (REIT) own the majority of the premier locations.

LPTs are often keen to reduce their water and energy consumption as a result of market pressures. These may include the following:

- Customer demand – More than half of the world's 500 largest corporations issuing sustainability reports in 2005 say that they want to build and occupy real estate that reflects their values [3].
- Market attractiveness. The green building sector is one of the fastest growing sectors within the commercial property sector.
- Pressure from institutional investors such as pension (Superannuation) funds.
- Regulatory pressure – Many governments are requiring that building managers and owners have well-developed water and energy management plans.
- Reporting requirements such as in the European VfU reporting requirements for the finance, banking and insurance sectors and Global Reporting Initiative (GRI).
- Inclusion in Sustainability indices such as the Dow Jones Sustainability Index and RepuTex.
- Competitive advantage – to have a green portfolio and this requires that the building are rated according to industry rating tools such as The Leadership in Energy and Environmental Design (LEED) Green Building Rating System™ or National Australian Built Environment Rating System (NABERS).
- To minimise current and future costs, liabilities and risks.

The average breakdown of water usage in an office complex is given in Figure 11.1

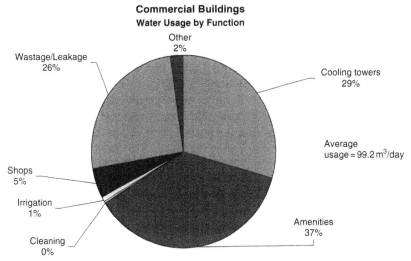

Figure 11.1   Water-usage breakdown in an average office building.

Courtesy of Sydney Water

The chart in Figure 11.1 shows that an average office building's water use can be minimised by one-third by arresting leaks. Cooling towers account for 29% of water usage. There are opportunities to save water in cooling towers and in amenity blocks, where the majority of water leaks occur.

11.2.2 Water-Usage Benchmarks

Water use in office buildings and consequently benchmarks are dependent on a number of factors:

- square area
- number of employees or tenants
- number and type of retail facilities
- type of water using facilities and efficiency
- climatic conditions.

Commonly used indicators to benchmark water use in commercial buildings are

- Water usage per square area – $m^3/m^2/yr$ (USgal./ft^2/yr)
- Water usage per employee/day – L/employee/day (US gal./employee/day)
- Water usage per annum per employee – m^3/yr/employee (Thames Water, UK).

In Australia, Sydney Water developed a best practice water use benchmark of $0.8\,m^3/m^2/yr$ (19.63 US gal./ft^2/yr) – based on over 30 comprehensive audits in large office buildings. This translates to $22\,m^3$/day for $10\,000\,m^2$ of floor space. However, this figure is only applicable if there are cooling towers in the building. If the building has air-cooled chillers, then the water use will be less. A study conducted by the Australian Department of Environment and Heritage [4] has identified that water usage in office buildings within Australian capital cities can vary from 0.4 to $3\,m^3/m^2/yr$ with a median of $0.91\,m^3/m^2/yr$. Using statistical analysis it also concluded that the most reliable indicator was $m^3/m^2/yr$ (after correction for cooling degree days which takes into account the heat rejection via cooling towers in warmer climates) rather than occupancy-based benchmarks.

In the United Kingdom, the government took the initiative (under its Watermark Program) to set a target of reducing water usage in government departments from $11\,m^3$/person/yr to $7.7\,m^3$/person/yr by 2004. For new buildings (built after 2002) the benchmark is set at $7\,m^3$/person/yr [5]. A best practice figure of $6.4\,m^3$/person/yr has been established. Thames Water has also established benchmarks for office buildings that ranges from $6.8\,m^3$/person/yr for buildings larger than $1000\,m^2$ to $4.4\,m^3$/person/yr for buildings less than $1000\,m^2$, and additional 1.5 to be added if there are catering facilities available [6].

Table 11.1   Environmental performance benchmarks for corporate ecology

	Water usage	Water usage per area*	Electricity Usage	Heating Energy
	Litre/employee/day	m³/m²/yr	kWh/employee	kWh/m²
Ellipson Best Practice Benchmark	47	0.7	3,706	119
Ellipson Standard Practice Benchmark	104	1.7	6,955	135
The best company performance	30	0.5	2,562	61
The average performance	104	1.7	6,955	135
The worst company performance	135	2.2	10,228	289

Adapted from Muller K. Sturm A. *Benchmarking Corporate Ecology of the Finance and Insurance Industry in Switzerland, Germany and Austria.*
* Based on UK average office space per employee of 16.3 m².

According to a study conducted by the University of NSW [7] in Australia, the average office space per person is 20.6 m², while in the UK it is 16.3 m².

On this basis the Water Mark target of 6.4 m³/person/yr equates to 0.4 kL/m²/yr.

In using the square area figure, care needs to be taken to ensure that the reported area is nett or gross surface area. In the United Kingdom and in Australia, office space is reported in Nett Lettable Area (NLA) otherwise known as nett internal area.

In Europe, the Finance and Insurance sector in Germany, Switzerland and Austria publicly reports environmental benchmarks under the VfU Guidelines. These are reported in L/employee/day. Researchers Muller and Sturm of Ellipson [8] have used this data to calculate water and other environmental best practice indicators for this sector. Table 11.1 shows water and energy usage per employee and if the average UK office space is used as a guide the expected water usage per m².

11.2.2.1   Energy Consumption

As a group office buildings are the largest energy user within the commercial property sector. In a typical office building lighting, heating and cooling represent more than 50% of total energy use. Office buildings in the United Kingdom (with air conditioning) will consume the following [9]:

Total annual consumption MJ/m²/yr	Good	Typical
Gas or oil	349	641
Electricity	461	814
Total	810	1455

Therefore water conservation strategies need to be geared towards reducing energy consumption which will result in a stronger business case.

11.2.2.2 Shopping Centres

Shopping centres play a major role in the OECD countries economies. For instance, in Australia there are 1338 shopping centres ranging from large regional centres of more than 100 000 sq m of retail space and generating sales of around A$500 million a year down to smaller, supermarket-based centres of around 5000 sq m generating sales around A$30 million. Cumulatively they generate $51 billion in retail sales each year [10].

From Table 11.2 it is evident that the trend in shopping centres is to build larger and larger shopping malls resembling mini-cities with thousands of shops, catering and other facilities. Eight of the ten largest malls in the world were in Asia by early 2007 and several more mega-malls in China and the United Arab Emirates are under construction. It is predicted that within the next few years, seven of the ten largest shopping malls in the world will be in China alone. The Australian company Westfield Group is the largest shopping mall owner today.

The results of six water audits conducted by Sydney Water, in six shopping malls (with an average gross lettable area between 23 000 and 116 000 m^2 (0.25–1.25 million ft^2)), and with an average water consumption of 120–660 m^3/day (32 to 127 thousand US gal./day), identified that leakage accounted for (on average) 32%.

Water usage in shopping centres is shown in Figure 11.2 with most of the leakage occurring in restrooms and cooling towers.

Table 11.2 Largest shopping centres in the world sorted by gross lettable area*

Mall	Location	Size (millions of square feet)	Size (square meters)
South China Mall	Dongguan, China	7.1	660,000
Aeromall (under construction)	Caracas, Venezuela	6.7	620,000
Golden Resources Mall	Beijing, China	6	560,000
Central World	Bangkok, Thailand	5.8	550,000
Mall of Arabia (under construction)	Dubai UAE	5.6	520,240
Seacon Square	Bangkok, Thailand	5.5	500,000
Runwal Arcade	Mumbai, India	4.25	425,000
SM Mall of Asia	Manila, Philippines	4.2	386,000
West Edmonton Mall	Edmonton, Alberta, Canada	3.8	350,000
SM Megamall	Manila, Philippines	3.6	332,000

Source: Eastern Connecticut State University, USA.
*Gross lettable area (GLA) is the total area of floor space leased for retail shops, consumer services, and entertainment, including restaurants. GLA is less than total area.

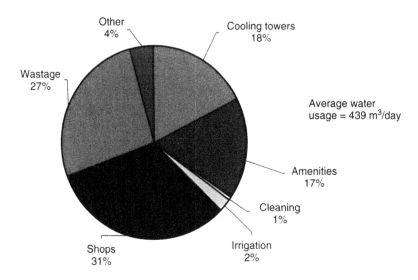

Figure 11.2   Breakdown of water usage in shopping centres

Courtesy of Sydney Water

From an energy-consumption perspective, in a typical retail building in the United States, lighting, cooling, and heating represent 69 and 84% of total energy use depending on the climatic conditions. Out of this, lighting accounts for 42–61% of consumption. Cooling represents 5–24% of total energy consumption and water heating varies from 1 to 7% [11].

11.2.3   Water-Saving Opportunities

Given that leakage accounts for a large part of water wastage, water efficiency can be improved with some easy steps before considering complex solutions.
   Here are some suggestions.

1.   Monitor water usage per square area per annum or per employee per day.
2.   Carry out a site-wide assessment of other sub-meters available on site. Ensure that these are in working order.
3.   Repair leaks.
4.   Inspect all toilets and replace gravity type toilets with dual flush 6/3 L, 4.5/3 L or high-efficiency toilets, pressure-assisted toilets as discussed in Chapter 10. The reduction in water usage from a 12 L/flush to 4 L/flush using a dual-flush toilet is 51 L/day [12].
5.   Inspect urinals. If they are the 'flush and fill' type, replace them immediately. Check timing cycles and volumes discharged. The calculation below shows how much of water (and dollars) go down the drain.

- Assume the cistern is a 'flush and fill' type which fills every 2 minutes. The capacity of the cistern is 6 L and the number of persons using the urinal is 50/day.
- The water consumption is 1576 m³/yr (416 thousand US gal./yr). This would cost \$2/m³–\$3153/urinal. Now assume that there are at least 10 of these in the building. The cost and water wastage now becomes significant. By installing a sensor that operates the urinal (based on movement) the water usage could be reduced to 0.2 m³/day (52 US gal./day) – easily justifying the cost of replacement.

6. Replace showers with water-efficient models (9 L/min) maximum.

   An office building with a floor area of 30,000 m² has 2000 employees (based on 15 m²/person). Ten per cent use the showers daily for an average of 8 minutes. The office is used for 5 days/week.

	Conventional	Water efficient showers
Water usage	15 L/min	9 L/min
Number of showers/day	200	200
Total water usage kL/day	24	14.4
Annual usage, kL	6,240	3,744
Total saving, kL		**2,496**
Total saving, %		**40.0%**

7. Install water-efficient aerators in taps. Typical taps discharge 15–18 L/minute. An ultra low water-efficient aerator has a flow rate of 1.7 L/minute. At the very least, a 6 L/minute should be the maximum allowable.

   For hand washing a 1.7 L/minute pressure-compensating flow controllers have proven to be adequate. Pressure compensation is essential to ensure that all taps have the same flow rate. Lack of pressure compensation may lead to an imbalance in water flow rates in fixtures, even though they are connected to the same pipe. For more details, refer to Chapter 10.

8. Convert once-through cooling water systems to open recirculating cooling water systems or reuse the water elsewhere. Refer to Chapter 5.

9. Consider air-cooled or hybrid cooling systems when replacing cooling towers Refer to Chapter 6. Hybrid systems save energy and water.

10. Install conductivity controllers in the cooling tower and boiler blowdowns to ensure that the correct conductivities are maintained and blowdown frequencies are not too frequent. Refer to Chapter 5.

11. Optimise usage of steam boilers and minimise blowdown as per Chapter 7.
12. Use trigger-operated nozzles on all hoses. Details are given in Chapter 12.
13. For irrigation of lawns and ornamental plants, use drip irrigation.
14. Check sprinkler systems and timing devices regularly to ensure that they operate properly.
15. Capture rain water if possible and reuse for irrigation, cooling tower makeup and toilet flushing.
16. Connect the water meters to the building management systems.
17. Capture condensate from air-handling units and reuse it for toilet flushing or as cooling tower makeup.
18. In kitchens, use pre-rinse spray valves for cleaning dishes before loading the dishwasher.
19. In shopping malls, individually meter the large retailers and issue bills according to actual water usage.
20. Cooling water treatment contracts to include water efficiency and penalty clauses for poor performance.
21. When testing fire water pumps consider reusing this water.
22. Only operate water features during working hours.
23. Initiate a tenant and employee awareness campaign. Contact the local water authority for assistance.
24. Request suggestions from employees.

Case Study: Jessie Street Centre Parramatta, Sydney.

Jessie Street Centre in Parramatta is a multi-storey office block in Parramatta, Sydney.

Sydney Water commissioned an audit and achieved water savings of $24\,000\,m^3$ in 6 months. The following initiatives were carried out to realise these savings:

1. Continuous flushing urinals were replaced with sensor-operated ones.
2. The irrigation of the building's garden was adjusted to eliminate excess watering.
3. All 324 toilets were modified to reduce their flush volumes.
4. A monitoring programme was put in place.
5. Initial capital outlay $40\,500. Cost savings $95\,000. Payback – 6 months.

Adapted from: *Sydney Water Fact Sheet – Jessie St Centre, Parramatta.*

11.3   Hospitals

Hospitals are large users of water and energy and emit a significant amount of greenhouse gases. It is also one of the largest employers in a country. For instance, in Australia nearly 220 000 people are employed in 1052 public and private hospitals [13,14].

Ageing populations, high cost of providing in-patient care in a climate of reduced budgets, provide the business case for hospital administrators to champion water- and energy-efficiency best practice and thus reduce these costs. According to the American College of Healthcare Executives, 67% of Chief Executive Officers list financial challenges as their number one concern. Every dollar saved in minimising water and energy usage is another dollar for patient care and a win for the environment.

Increasingly hospitals are also responsible and accountable to a range of stakeholders who use, render, regulate and benefit from hospital services. Therefore it is in the hospitals' best interests to promote best practice resource efficiency and communicate this to all stakeholders to minimise costs and budget cuts.

There is a direct link between healing the individual and healing this planet... We will not have healthy individuals, healthy families, and healthy communities if we do not have clean air, clean water, and healthy soil.

Lloyd Dean, MA CEO Catholic Healthcare West, October, 2000.

Source: Energy Star. Making the Business Case for Energy Management Healthcare.

Hospitals can be classified as in-patient, day-hospitals and community-based hospitals. The water usage in hospitals depends on

- the number of patients being treated
- number of beds
- type of hospital (teaching and research or community hospital)
- age of the hospital
- type of medical treatment provided
- and the types of facilities available in these hospitals – such as the presence of on-site laundries.

The major water-consuming areas in hospitals are

- domestic water use for toilets, wash basins and urinals
- medical equipment, X-ray processors, dialysis machines, sterilisers, autoclaves and vacuum systems
- on-site laundry

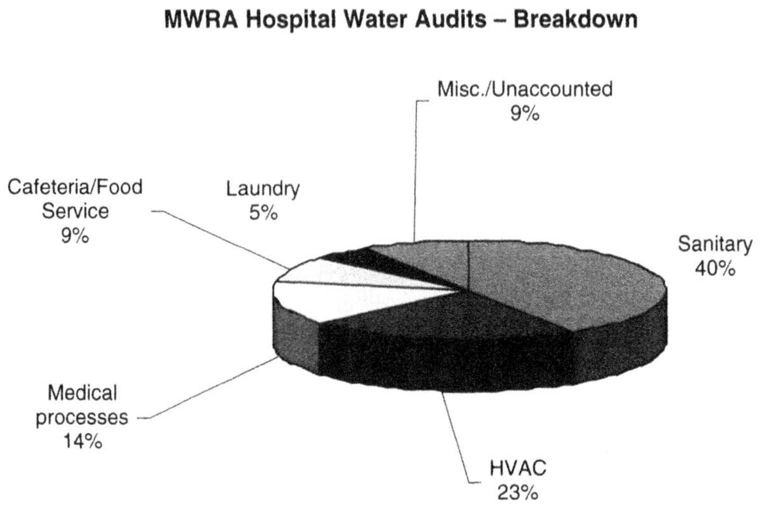

MWRA Hospital Water Audits – Breakdown

Misc./Unaccounted
9%

Cafeteria/Food
Service
9%

Laundry
5%

Sanitary
40%

Medical
processes
14%

HVAC
23%

Figure 11.3   Average water usage in seven hospitals [15]

Courtesy of Massachusetts Water Resources Authority.

- laboratories
- air-conditioning and refrigeration systems
- reverse osmosis systems for dialysis machines
- boilers or hot water generators
- kitchens and cafeterias
- landscaping
- leakage.

Figure 11.3 shows the average water used in hospitals within the Metropolitan Boston area, derived from the water audits of seven hospitals, ranging from 138 to 550 bed capacities. These had a water usage ranging from 155 to 700 m³/day (15–67.2 million US gal./yr) [15].

11.3.1   Benchmarking Water Usage

Two common water usage benchmarks for hospitals are

1. Water usage per area – m³/m² (US gal./ft²)
2. Litres/bed/day (US gal./bed/day)

Table 11.3 shows water usage guidelines developed for UK hospitals [16, 17]. Table 11.4 shows the U.S. Department of Energy's rough estimates of water usage for hospitals [18].

11.3.2 Benchmarking Energy Consumption

All hospitals are unique in design and the services they provide and also have special requirements unlike other buildings. For instance, indoor temperatures may be slightly higher to maintain patient comfort levels. Seventy-five per cent of energy consumption in hospitals is for lighting, ventilation, space heating and water heating.

A 1997 study conducted by the European organisation, Centre for the Analysis and Dissemination of Demonstrated Energy Technologies (CADDET), concluded that there was a large discrepancy in electricity consumption amongst the six countries that participated in the survey [19]. For instance, electricity consumption per bed varies from 5.1 MWh (Italy) to 28.1 MWh (Australia), with an average consumption of 16.1 MWh (58 GJ/bed/yr). This is explained by a variety of factors such as climatic conditions, prices, age of equipment and insulation in buildings that can impact on electricity consumption. On the other hand, thermal energy consumption per bed is more uniform, varying between 23.3 MWh (Italy) and 42.8 MWh (Canada) with an average of 33.9 MWh (122 GJ/bed/yr).

However, similar to water consumption benchmarks there are large variations between acute, teaching and small hospitals.

Electricity consumption per floor area varies between 61 kWh/m^2/yr (Switzerland) and 339 kWh/m^2/yr (Canada) with an average consumption

Table 11.3 Water usage guidelines for hospitals in the United Kingdom by usage per patient bed-days

Type of hospital	Litres per patient bed-day			
	Very poor	Poor	Average	Good
Acute > 100 beds	1138	711–1137	531–700	<530
Long stay > 25,000 patient days per year	690	412–689	331–411	<330
Long stay < 25,000 patient days per year	380	298–379	218–297	<217

Courtesy of Thames Water, UK.

Table 11.4 Water Usage Guidelines for hospitals in the United States

Unit	Range	Typical
US gal/bed	80–150	120
L/bed	303–568	454
US gal/employee	5–15	10
L/employee	19–154	38

Adapted from U.S. Department of Energy – Energy Efficiency and Renewable Energy – Federal Energy Management Program – Water use indices [18].

of 145 kWh (522 MJ/m²/yr). Thermal consumption varies between 168 kWh/m²/yr (Sweden) and 690 kWh/m²/yr (USA) with an average of 367 kWh/m²/yr (1321 MJ/m²/yr).

11.3.3   Water Conservation Opportunities

Water conservation opportunities in hospitals are similar to those mentioned previously for commercial office buildings and shopping centres. Some specifics relating to hospitals are given below.

11.3.3.1   Monitor Leakage

Many hospitals have sprawling infrastructure and leakage in underground pipes is common. Sydney Water monitored 15 hospitals' main meters electronically for a 2-week period. It showed that close to 50% of hospital water usage was due to water leaks – known as base flow. Lack of adequate monitoring equipment makes it hard to notice these leaks. Given the number of toilets, urinals and wash basins generally available in these facilities, having a good leakage minimisation programme will be a good starting point to reduce on-site water usage.

Rule of Thumb for checking for leaks

If the base flow rate is significantly greater than 10% of the peak flow rate, then this could be because of leakage.

Case Study: Westmead hospital, Sydney, Australia

Established in 1978, it is one of the largest specialised referral hospital in Australia. It services a population of over 1.5 million people and occupies over 130 000 m² (1.4 million ft²).

Sydney Water with the assistance of the Department of Commerce identified leaks of 50 000 m³/yr which was a saving of A$ 80 000/year. The hospital has set itself a target of reducing water consumption by 25% through a range of initiatives such as:

Installing water efficient showers and taps including 5800 flow restrictors, replacing flush valves on all toilets, with an expected savings potential of 50 000 m³/yr and a monetary saving of A$ 100 000/year.

Other initiatives under consideration include use of ground water as make-up water in cooling towers which could save another $100 000/yr.

Since 2001 Westmead hospital has reduced overall water consumption from 420 000 m³/yr to 333 000 m³/yr.

Adapted from: *Sydney Water. Hospitals Saving Lives and Water. The Conserver.* Issue 6. December 2004 [20].

11.3.3.2 Air-conditioning and Refrigeration Systems

Air-conditioning systems in hospitals in comparison to other building types have an added complexity for medical requirements that

- the air movement needs to be restricted between buildings to prevent the spread of disease causing germs
- stringent requirements for ventilation and air filtration. For an example, air changes can be as high as 20/hr in operating theatres.
- different temperatures and humidity requirements for different hospital areas.

The water conservation opportunities are similar to those found elsewhere. Refer to Chapter 5 and 6.

Common actions are

- eliminate once-through cooling
- eliminate leaks and overflows
- minimise blowdown and drift
- recover condensate from air-handling units (given the high rate air changes per hour)
- use alternative supply sources
- replace old or inefficient cooling towers with hybrid cooling towers
- investigate heat recovery options in HVAC units.

11.3.3.3 Steam Systems

Steam is used extensively in hospitals for heating, to sterilise medical equipment, in absorption chillers, in laundries, for food preparation and so on. Maximising return condensate, minimising boiler blowdown and having a well-designed maintenance programme will reduce water and gas usage and costs. For more details, refer to Chapter 7.

11.3.3.4 Taps, Toilets and Urinals

Taps refer to Chapter 10 for a discussion on flow restrictors and pressure-compensating aerators. Replace flush and fill urinals with water-efficient models. Replace single flush gravity toilets with dual-flush models as discussed in Chapter 10.

In hospitals apart from gravity flush toilets, flush valve-operated toilets are common. The 'flushometer valve' is directly connected to the water supply plumbing. The valve controls the quantity of water released over time by each flush. These valves need a minimum water pressure of 200–237 kPa (30–35 psi) which is significantly higher than the gravity-operated ones which only require a pressure of 4 kPa (0.6 psi) to work effectively. Flush volumes are nominally rated at 6 L/flush (1.6 US gpf) or 13 L/flush (3.4 US gpf). The flush volumes do not change with changes in pressure according to a study conducted by Veritec Consulting [21]. The study found that the flush volumes can be changed in some models from 6 L to as much as 18 L by replacing the diaphragms and piston valves, adjusting the adjustment screw, changing internal O-ring or control stops.

11.3.3.5   Food Preparation

Hospitals have large kitchens which are high water-using areas. Water is used for preparing food and washing equipment and containers. These have been discussed in Chapter 10.

Water savings can be made by

- installing pre-rinse spray valves
- inspecting dishwashers and ice-making machines
- considering foot-operated taps for wash basins
- not defrosting food done under flowing water
- minimising water usage to flush out potato peelings
- using a broom first before washing of floors
- using a hose with a water-efficient trigger gun.

Case Study: John Umstead Hospital, North Carolina, U.S.A

John Umstead Hospital in Butner, North Carolina is a 593-bed psychiatric hospital for patients older than six years.

A water conservation programme identified that it had five ice-making machines in the kitchen and four of them had single-pass cooling, consuming a total of 1623 m³/yr (429 000 US gal./yr). The hospital staff consolidated the ice-making machines and piped it into an existing cooling tower, saving that much water.

Adapted from: *North Carolina Division of Pollution Prevention and Assistance Case Study – John Umstead Hospital.*

11.3.3.6   In-house Laundries

Hospitals have either in-house laundries or a central laundry that provides a service to a number of hospitals within its geographical locality. Refer to Chapter 15.

11.3.3.7 Medical Equipment

i) X-ray processing equipment

Medical facilities commonly have X-ray film processors operating 365 days of the year. Most have more than one processor. X-ray processing equipment uses water to rinse chemicals from the film before it reaches the dryer section of the machine and to cool the equipment. The rinse section receives a constant supply of water. The published water usage rate is estimated to be around 0.94–9.4 L/minute (0.25–2.5 US gpm) [22]. In older machines this water is rarely captured and reused. Moreover, in practice, it has been found that the water flow rates are one-and-a-half times (to twice) this amount. They therefore have the potential to waste large quantities of water as seen by the case study.

The X-ray film development process in brief is as follows. The process consist of developing, fixing, washing and drying of the film. The X-ray film goes to a tank which holds the developer chemicals which is maintained at a constant temperature of 35° C(95° F). The film then passes through rolls to remove excess water before being exposed to the fixer chemicals. The fixation process stops the development chemicals from over-exposing, after which the film is washed or rinsed with flowing water to remove excess chemicals. After the excess water is removed the film is then dried. The water-saving opportunities are

- Install a sub-meter to monitor the supply line flow received by the rinse water.
- Install a control valve to adjust the flow rate to the manufacturer's recommended flow rate.
- Install a solenoid valve to shut water supply to the rinse tank and cooling water pipes when the unit is not in use.
- Install squeegees to squeeze the maximum amount of liquid to minimise carryover as the film travels from one tank to another. (This reduces wash water requirements.)
- Recycle the rinse water. Water Saver/Plus™ units have been installed in California in many older units with considerable success. These units costs US$4000–5000 and operating costs are US$1300 [22].
- Shut off all units not in service.
- Replace old machines with digital imaging processors. They provide better images and at lower unit cost is more cost effective over the long term and does not use water.

Case Study: Water Saver/Plus™

Irvine Ranch Water District in California installed 38 Water Saver/Plus™ units in 7 hospitals to recirculate the water from the X-Ray machines.

Water Saver/Plus™ consists of a small reservoir with a capacity of 56 L (15 US gal.), a pump and an algaecide dispenser. A water meter and a data logger were installed prior to the units were installed to log water usage data for one month before and after installation of the Water Saver/Plus™. A timer releases a set amount of fresh water per hour for temperature control. The results showed that the device saved $4.5\,m^3$/day (1183 USgal./day) or 1.63 ML/yr (1.33 acre ft/yr) – a 97% reduction in water usage.

Adapted from: *Water Conservation News* January 2003 [23].

ii)   Steam sterilisers and autoclaves
Steam sterilisers (also known as autoclaves or pre-vacuum sterilisers) are used in hospitals and other institutions requiring sterilisation of medical equipment in the manufacture of pharmaceutical products – for the total destruction of microorganisms. The most common type of steriliser is the steam steriliser.

Steam sterilisers use water for jacket cooling and for generating a vacuum using steam ejectors (vacuum type sterilisers).

The process is as follows: Medical equipment, after going through a washing process, is placed in the steriliser. The sterilisation process consists of removing air to create a vacuum seal and injecting low-pressure steam at 134° C (273° F) into the chamber for about 4 minutes. After sterilisation the vacuum seal is broken to start the drying process. The whole process takes usually 35 minutes to complete. The steam condensate leaves the steriliser at around 80° C (176° F). Before discharging this water to the sewer, it needs to be cooled below 38° C (100° F) to comply with trade waste regulations (since water authorities prohibit the discharge of water at temperatures above 38° C (100° F)). Steriliser jackets are also kept warm by maintaining a flow steam to the chamber to minimise steam condensation on the walls and minimise start up times. This water is sent to the drain. At 11 L per minute the water wastage during one cycle is around 418 L.

Water conservation opportunities are

*   Jacket and chamber condensate cooling modification [22]
    Steam that is condensed from the jacket during the 'standby' mode is discharged to the sewer. Therefore, cool potable water is typically used to mix with the hot water before discharging to the sewer. No attempt is used to conserve this water and it runs continuously. Flow rates of cool potable water wasted are in the range of 11 L/minute.

The condensate can be modified to collect the jacket water in a tank, and letting the ambient air circulating around the tank to cool it first before discharging to the sewer. When the temperature rises above the legally permissible temperature of discharge to the sewer, a thermostatically actuated valve opens to inject the minimum quantity of potable water which is added to cool the condensate temperature below the set point. The whole unit is small enough to fit in space-constrained areas. This measure has the potential to reduce the water usage by 85%.

- Ejector water modification
  Potable water is used as the driving force to create a vacuum in the sterilisation chamber. The potable water is traditionally wasted. The modification consists of collecting this water in a tank and then using it in place of fresh water as the driving force for the vacuum. The principle is based on the ambient air cooling the water in the tank below 49°C (120°F). When the temperature exceeds the set point, cool fresh water is used to cool the tank water below the set point.
- Use chilled water for cooling of hot condensate, liquid ring vacuum pump or steam ejector water and reusing it. The chilled water is circulated in a closed cooling coil, cooling the condensate before discharging it to the sewer [24]. The relatively warm chilled water returns to the chiller. Read the case study below for more details.
- Shut off all units not in service.

Case Study: Western Health Victoria retrofits steam sterilisers and reaps a quick pay back

Western Health Victoria retrofitted 9 of their steam sterilisers which included a chilled water loop to cool the condensate to save 20 235 m³/yr at a cost of A$60 400. The payback was 2.4 yrs.

Flow rate of cooling water	22 Lpm
Time for each cycle	35 minutes
Number of cycles per day per unit	8
Number of units	9
Quantity of water used and saved per year	20 236 m³
Cost of water and sewerage	A$24 950
Capital cost	A$60 400
Payback	2.4 yrs

Adapted from: *Water Saving Sterilisers, Western Health, Smart Water Fund, Melbourne, Victoria* [24].

iii) Medical pumps and liquid ring vacuum pumps

Liquid ring vacuum pumps and steam jet ejectors are used in hospitals to create a vacuum in sterilizers and in laboratories. These can waste significant quantities of water. Details on vacuum pumps are given in Chapter 13. Solutions are to recirculate the cooling water or replace the pump with an oil ring or a dry vacuum pump. The case study given below illustrates the point.

> **Case Study: Royal North Shore Hospital (RNS), Sydney replaces liquid ring vacuum pumps and reaps the rewards**
>
> RNS is a 740 bed acute general teaching hospital. The hospital received a subsidy of A$104 025 from Sydney Water to replace their medical suction and laboratory vacuum pumps with dry running pumps. Since then RNS is saving 100 000 litres of drinking water daily. The annual savings in water charges alone exceeded $60 000. The project had a simple payback of less than two years.
>
> Adapted from: *Sydney Water. The Conserver issues August 2005 and December 2005.*

iv) Hemodialysis units

Water used in hemodialysis units is purified beforehand in reverse osmosis systems (RO). The reject water from the RO system is usually sent to waste. If the TDS is not excessive this water can be used as cooling tower makeup or for flushing of toilets. Consult the cooling water chemical treatment supplier for advice. For details on RO systems refer to Chapter 8 and cooling systems refer to Chapter 5.

> **Case Study**
>
> East Kent & Canterbury Hospital is one of the three main hospitals in the East Kent Hospitals Trust. The Trust has a duty to promote the efficient use of resources. It identified an opportunity to save water and money at the Renal Unit at East Kent & Canterbury Hospital.
>
> The project involved recycling the wastewater that had been sent to drain by the Reverse Osmosis (RO) plant. The RO plant is used to separate specific calcium ions in the water and purify it for use by the renal unit. The new initiative diverted water to a holding tank via pipe work and then subsequently to a larger redundant tank in a separate plant room. This water was then re-used to flush all the toilets and urinals in the main operating theatres and the accident centre areas. Mains water is used as a top up when required.
>
> The initial set-up costs included a feasibility study, materials (piping and one holding tank) and installation costs. The annual maintenance costs of the system are minimal.

> The Trust aimed to recover the costs of the project in less than three years. They based this on their calculations for wastewater passing to drain and subsequent savings on WC/urinal usage at the prevailing water and sewerage rates. This target was achieved with a 38% reduction in the use of mains water and on-going annual financial savings of £7000.
>
> *Courtesy of Environment Agency UK.*

11.3.3.8   Increase Staff Awareness

Given the large number of employees, increasing staff awareness will play a significant factor in reducing water usage.

11.3.3.9   Floor cleaning

Floor cleaning is an essential part of a patient care program to minimise the spread of pathogens. In conventional wet mopping of floors, cotton mops are used. A disinfectant is added to the water and after every 2–3 rooms the water is discarded. At the end of the shift the mops are sent to the laundry for washing and drying.

Microfibre mops are considered to be more water efficient than the cotton mops. Constructed from nylon and polyester fibres, they are approximately 1/16 the thickness of a human hair. The density of the fibres enables it to hold six times its weight in water. As a result for a hundred room hospital it only requires 19 L (5 US gal.) of water as against 397 L (105 US gal.). The microfibres are split and they also carry a positive charge which enables them to attract dust particles and thereby increasing the cleaning efficiency. The mops also can withstand 300–500 washings as against 55 for cotton mops.

According to a study carried out by the University of California Davis Medical Cente, the microfibre mops are about 5–10% cheaper than using cotton mops. The studies concluded that microfibre mops reduced water and chemical usage by 95%; Lifetime cost savings by 60%; Labour saving by 20% and were ergonomically better for the janitors due to less lifting [25, 26].

11.4   Correctional Centres

Prison populations are increasing with newer prisons being built by the public and private sector. For example, in the United Kingdom there are 135 prisons with a current prisoner population in excess of 65 000 whereas in 1985 there were only 41 500 prisoners [27]. Given the increase in prisoner populations (in most countries) both energy and water usage is increasing the cost of maintaining prisoners, stretching the limited budgets of state and local governments.

Prisons are categorised as high security, local, open, women's or juvenile offenders' institutions. The prison category influences the water usage. In high-security prisons usually there will be a higher staff /prisoner ratio. In New South Wales, Australia, the staff to prisoner ratio is around 2–3. Open prisons may have low occupancy levels during the day when prisoners work off-site. On the other hand, these may have other water-using facilities such as farms, dairies and vegetable canning works.

11.4.1 Water Usage Benchmarks

The water usage benchmarks for correctional centres are expressed as:

Water usage per floor area – $m^3/m^2/yr$
Water usage per prisoner   – $m^3/prisoner/yr$ or $m^3/prisoner/day$.

Due to the occupancy characteristics, water consumption in detention and correctional facilities is high. Kitchens, laundries and shower facilities require large volumes of hot water. Inmate areas require special plumbing fixtures to minimise vandalism and suicide attempts. These aspects are seen from the high water-usage figures per prisoner. The UK Watermark Project [27] has established the benchmarks shown in Table 11.5. An international comparison is given in Table 11.6.

11.4.2 Water Conservation Opportunities

Prisons are not dissimilar to other institutions, except that they have special requirements. With traditional plumbing fixtures prisoners have full control over lavatories, toilets and showers. This can lead to a variety of problems, such as intentional flooding through simultaneous flushing and constant

Table 11.5 Water consumption for UK prisons

Category	Typical m^3/prisoner/yr	Best Practice m^3/prisoner/yr
Prisons with laundry	143	115.3
Prisons without laundry	116.6	92.4

Source: Environment Agency UK.

Table 11.6 Water Consumption for prisons – international comparison

	NSW Australia	USA	UK
Typical water usage m^3/inmate/day	0.32–1.29 [29]	0.30–0.57*	0.32–0.39**

* Adapted from The US Federal Energy Management Program (FEMP) water indices data gives a figure 80–150 US gal./prisoner/day.
** Adapted from Table 11.6.

running of showers and lavatories. To overcome these problems as well as minimising water usage, more attention needs to be given to the planning aspects (as well as remote monitoring of fixtures) through innovative use of technology and the installation of purpose-built fixtures.

Some suggestions are given below.

- Install sub-meters.
- Data log main and sub-meters to identify water leakage. It is preferable that the dataloggers are connected to the building management system.
- Install systems that enable the shut off of individual or group plumbing fixtures in a cell (or cells) rather than shutting down the entire domestic water system.
- The main use of water is for toilet flushing and washing. Install movement-detection sensors to prevent wastage when the toilets are unoccupied.
- Conceal flush valves and infra-red detectors to prevent them being tampered with.
- Use electronic controls to prevent inmates wasting water by programming the number of flushes within a specified time frame. This also allows the staff to monitor the frequency of toilets usage. When inmates try to flush the toilet more than the specified number, the toilet is locked out. In one instance, a prison had a problem of trying to prevent inmates flushing sheets and clothing down toilets and blocking toilets. By monitoring the frequency of usage, staff are able to detect the problem in time and prevent flooding of cells.
- Use integrated sink and toilets with push button controls to minimise vandalism.
- Replace water guzzling toilets with 13 L/flush (3.5 US gpf) with vacuum toilets that use only 1.2–2 L/flush (0.3–0.5 US gpf). Vacuum toilets have long been used in the aviation and marine industry. There is anecdotal evidence to suggest that prisoners tend to flush toilets 15–20 times a day [30]. With a vacuum toilet the water used is only 40 L/day (10.6 US gal./day) as against a conventional toilet using 13 L per flush wasting 260 L/day (70 US gal./day). When the cost of water and sewage is considered this could result in considerable savings – as seen from the example below.

The other advantage of vacuum toilets is that if an object was too large to pass through the bowl into the vacuum tubes, computer sensors would shut down the affected toilet and alert a system operator to the problem, as well as pinpoint the culprit. During cell inspections for contraband, security staff can use the computer-control network to shut down toilets in specific prison cells or to shut down the entire system, making it difficult for prisoners to flush away contraband.

Worked example

Number of inmates	500
Number of toilets	500
Number of average flushes per day	15
Water used with conventional toilet having 13 L/flush	$13 \times 15 \times 500 = 97,500\,\text{L/day} = 97.5\,\text{m}^3/\text{day}$
Water used with vacuum toilet	$2 \times 10 \times 500 = 10,000\,\text{L/day} = 10\,\text{m}^3/\text{day}$
Cost saving at $4/kL for water and sewage	$\$4 \times (97.5 - 10) \times 365 = \$127,750$
Plus other productivity savings from reduced maintenance over a 10 year period.	$\$127,750 \times 10 \times 1.50 = \$1.9$ million

The other benefits of vacuum systems are that the pipe diameters are smaller than for conventional gravity systems and therefore can be easily routed around beams and ventilation ducts.

- Install vandal proof fixtures (such as aerator spray heads) that are secured with a key. Shower heads that do not protrude from the walls are therefore difficult to tamper with.
- Follow other recommendations similar to hospitals for the kitchen and laundry.
- Consolidate individual laundries with one central laundry.
- Practice smart irrigation practices. Refer to Chapter 10.

References

[1] Gleick P.H., et al. *Waste Not, Want Not: The Potential for Urban Water Conservation in California*. Pacific Institute. November 2003.
[2] World Business Council for Sustainable Development. *Zero net energy building project recruits nine more international companies*. www.wbcsd.org. 5 September 2006.
[3] World Business Council for Sustainable Development. Commercial Demand for Green Development Grows. www.wbcsd.org. 7 August 2006.

[4]   Dr. Bannister P., Munzinger M. and Bloomfield C. *Water Benchmarks for Offices and Public Buildings*. Department of Environment and Heritage. Canberra Australia. www.deh.gov.au. September 2005.

[5]   Department for Environment, Food and Rural Affairs. *Framework for Sustainable Development on the Government Estate*. www.sustainable-development.gov.uk/publications/report2002/part1/.

[6]   Thames Water. Office Water efficiency. www.thameswater.co.uk.

[7]   Warren C.M.J. *A Comparison of Office Space Use in Australia and the UK*. A Paper presented at the EuroFM Rotterdam. www.rics.org. au. May 2003.

[8]   Muller K. and Sturm A. *Benchmarking Corporate Ecology of the Finance and Insurance Industry in Switzerland, Germany and Austria*. Ellipson AG. www.ellipson.com. August 2000.

[9]   Energy Efficiency Best Practice Programme. Good Practice Guide 285 – *What will energy efficiency do for your business*. Waterford, UK. March 2000.

[10] Shopping Centre Council of Australia. Vital Statistics. www.propertyoz.com.au.

[11] Platt Research & Consulting. Managing Energy Costs in Retail Buildings. Boulder Colorado. 2003.

[12] Department of Environment and Heritage. *Water Efficiency Labelling and Standards Scheme – fact and figures*. www.deh.gov.au.

[13] Parliament of Australia. *Hospitals in Australia*. www.aph.gov.au.

[14] Australian Private Hospitals Association. *Facts and Figures*. www.apha.org.au. 2006.

[15] Massachusetts Water Resources Authority. *Water use Case Study: Norwood Hospital*. www.mwra.state.ma.us/4water/html/bullet1.htm.

[16] Environment Agency. *Water Resources: How Much Water Should We be Using?* www.Environment-Agency.gov.uk.

[17] Thames Water. *Water Efficient Hospitals*. www.thameswater.co.uk.

[18] U.S. Department of Energy. Energy Efficiency and Renewable Energy. Federal Energy Management Program. *Water Use Indices*. www.eere.energy.gov/femp/technologies/water_useindices.cfm.

[19] CADDET. *Energy Efficiency in Hospitals*. Sittard, The Netherlands. 1997.

[20] Sydney Water. *Hospitals Saving Lives and Water*. The Conserver. Issue 6. December 2004.

[21] Veritec Consulting Inc. *Testing of Popular Flushometer Valve/Bowl Combinations – Final Report*. Mississauga, Ontario. May 2005.

[22] Koeller and Company. *A report on Potential Best Management Practices*. California Urban Water Conservation Council. Sacramento, California. August 2004.

[23] Gonzalez D. *Irvine Ranch's Water District's X- Ray Processor Retrofit Recirculates Water and Returns Savings*. Water Conservation News. Department of Water Resources. Sacramento, California. January 2003.

[24] Smart Water Fund. *Western Health: Water Saving Sterilisers.* www.smartwater.com.au. March 2005.

[25] US Environmental Protection Agency. *Using Microfiber Mops in Hospitals.* November 2002.

[26] Desa J., et al. *Are Microfiber Mops Beneficial for Hospitals?* University of Massachusetts, Lowell. 2003.

[27] Davis B. *New Regime for Prison Utilities.* HM prison Service UK. 2000.

[28] Environment Agency. *Water Resources: How much water should we be using?* www.Environment-Agency.gov.uk.

[29] NSW Department of Commerce. *Benchmarking Review 2003 – Water usage at Correctional Centres.* Sydney. July 2003.

[30] Hafner W. *Vacuum toilets Provide Solutions in Correctional Facilities.* PM Engineer. www.pmengineer.com. August 2000.

Chapter 12

Swimming Pools

12.1 Introduction

Councils, leisure and fitness centres, hotels and most motels have swimming pools.

Swimming pool water is an expensive commodity given the fact that a large body of water needs to be continually pumped, treated, filtered and backwashed and then heated to temperatures 26°C–30°C (79°F–86°F). In indoor pools the air needs to be ventilated. For this reason swimming pools consume two to three as much energy as an air-conditioned office building per square area. Table 12.1 shows the typical average energy use in sports centres with pools and dry sports centres in the UK [1].

If the values in Table 12.1 are compared against a typical air-conditioned office building in the UK, an office building consumes uses 226 kW/m² and a typical swimming pool uses 545–745 kW/m² of energy. It becomes evident the energy consumption of swimming pools is two to three times per square metre. Twenty-five per cent of the energy is used to heat pools to maintain these temperatures and another 53% of energy is used for space heating in indoor pools [1]. Therefore minimising water usage results in minimising energy usage and costs.

Figure 12.1 shows the typical breakdown of water usage in large public swimming pools. As the breakdown shows, retrofitting showerheads and minimising leakage will have a significant effect on reducing water use. Sydney Water data indicates that 33% of the water usage can be reduced by instituting good management practices [2].

12.1.1 Swimming Pool Benchmarks

Typical water usage for swimming pools in Sydney is around 70 L/visitor/day. This can be reduced to 40 L/visitor/day [2]. Table 12.2 gives the benchmarks for public swimming pools.

Table 12.1   Typical annual energy use (kWh/m²) [1]

	Sports centre with pool	
Good	Fair	Poor
<510	510–745	>745
	Dry sports centre	
Good	Fair	Poor
<290	290–410	>410

Courtesy of Energy Efficiency Best Practice Programme. Good Practice Guide 219 – *Energy Efficiency in Swimming Pools – for Centre Managers and Operators.* September 1997.

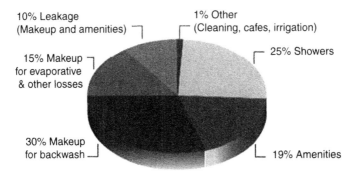

10% Leakage
(Makeup and amenities)

1% Other
(Cleaning, cafes, irrigation)

15% Makeup
for evaporative
& other losses

25% Showers

30% Makeup
for backwash

19% Amenities

Figure 12.1   Typical water usage breakdown in public swimming pools

Courtesy of Sydney Water. *Saving Water in Community Swimming Pools or Leisure Centres.* March 2005.

Table 12.2   Swimming pool benchmarks

Number of visitors	Benchmark
0–500	60 L/visitor/day
>500	40 L/visitor/day

Courtesy of Sydney Water. *Saving Water in Community Swimming Pools or Leisure Centres.* March 2005.

12.2   Water Conservation Opportunities

Water conservation opportunities for swimming pools consist of:

- reducing leakage
- installing water-efficient taps, showerheads and toilets
- reducing backwash frequency and time
- reducing pool evaporation.

These are discussed below.

12.2.1 Reducing Leakage

Hidden leaks are common in swimming pools. Leakage from cracks in tiles and more frequently from fixtures such as broken ball valves can waste a lot of water and can go undetected for a long time.

Sometimes leaks are mistaken for water losses from evaporation. To check whether the water loss is due to evaporation or a water leak, place a bucket of water on the top step of the pool and fill it with water to the pools water level. After a day if the water level in the pool is lower than the bucket, there probably is a leak in the pool structure or plumbing system.

To check whether the plumbing system is the cause of the leak, measure the water loss with the pump running for 24 hrs and again with the pump off. If more water is lost when the pump is running, the plumbing system is probably the cause.

To detect leaks in pool walls, a dye is added to the water and the flow of the dye is used to detect any cracks.

More sophisticated techniques consist of inserting television cameras into pipes to spot leaks. The camera delivers a clear picture on a video screen showing the problem while a transmitter shows the location of the leak.

Another method is to use a supersensitive microphone to detect leaks below concrete pool decking or in pool walls. By injecting air or inert gas into a pipe, then listening electronically for sounds of air or gas escaping, the technician can precisely locate the leak.

Case Study: Rooty Hill RSL Club saves A\$56 000/yr

Rooty Hill RSL club is one of Sydney's largest clubs with over 38 000 members. An audit revealed that the pool was using over $9\,m^3$/day – a third of the club's water consumption. A broken ball valve was sending water from the water flow trenches directly to the drains instead of returning to the pool. Fixing this single problem saved over $60\,m^3$/day.

By installing sub-meters and data loggers base flow can be detected in a timely manner. Refer to Chapters 3 and 4.

12.2.2 Water-Efficient Fixtures

The two previous chapters discussed the importance of installing water efficient fixtures in amenities blocks. In changing rooms it is common to have a long row of showers. Ensure that adequate flow balancing is present to minimise water wastage. Other suggestions are:

- To minimise water and energy wastage for hot water pipes, insulate pipes and locate the heaters close to the showers.
- Use a broom to wipe wetted areas.

- Lower the pool's water level by a few centimetres to reduce losses from splashing.
- Keep the pool and filters clean to reduce filter backwash frequency.
- Regularly check for leaks and cracks in pools and spas. Leaks can occur from cracks in tiles, suction and discharge side of pumps and filters.

12.2.3 Optimising Filter Backwash Cycles

Pool water is continually circulated to achieve the required levels of filtration and disinfection levels specified by health authorities. Swimming pool filters normally consist of pressure sand filters and coated mesh filters. Our focus is on pressurised filters which are the most common in large swimming pools. Pressure sand filters are housed in a cylindrical pressure vessel made of steel or moulded fibreglass. Refer to Chapter 8 for more details. There are predominantly two types of pressure sand filters:

1. Rapid sand filters
2. High rate sand filters

The difference lies in the flow rate. Rapid sand filters are limited to a flow of 200 L/m^2/min (5 US gal./ft^2/min). High rate sand filters have significantly higher flow rates of 600–1000 L/m^2/min (15–25 US gal./ft^2/min) and therefore suitable for space-constrained areas.

In the sand filters suspended solids are captured and in the process the filter media gets blocked with suspended solids. Consequently filter performance declines over time. This is indicated by an increase in the pressure drop between the inlet and the outlet of the filter. At a given set point the filter needs to be backwashed to remove the trapped particulate matter by running the pool water to waste. At the very minimum, filters need to be backwashed once a week. By doing so the backwash times are kept to a minimum (generally limited to 5–10 minutes) thus reducing the water used. Filter manufacturers specify the minimum backwash period. The end point of backwash cycle is when the water is clear.

When the pool water is used for backwashing it allows for fresh water to be added to the pool. This serves the purpose of diluting the chemical contaminants in the pool that are not removed by filtration such as dissolved solids, chloramines (formed due to interaction of urine with chlorine) and so on.

The example given below in Table 12.3 shows the impact of excessive backwashing on water costs.

It is possible to reuse the filter backwash water for irrigation as well as capture the heat contained in the backwash water by the use of heat exchangers. Water reuse will be dictated by the dissolved solids in the water. If the dissolved solids are high such as in salt water pools which may have a total dissolved solids content between 4000–7000 mg/L the water cannot be directly used for irrigation. Membrane filtration will be required to reduce

Table 12.3 Worked example – Cost of excessive backwash

	Backwash time 5 mins + 2 mins rinse	Backwash time 10 mins + 4 mins rinse
Number of filters	3	3
Area of filters	4.67 m²	4.67 m²
Water flow rate required to fluidise the sand bed	2.3 m³/min	2.3 m³/min
Backwash + rinse time	7 minutes	14 minutes
Water flow rates m³	48.3	96.6
Annual cost of water, wastewater discharge plus treatment at A\$3.00/m³	A\$52,889	A\$105,777

the TDS levels to acceptable levels. Refer to Chapter 8 for more details on membrane filtration. Contact the local environmental protection agency for acceptable limits.

Other recommendations are to carryout regular manual cleaning of the pool, skimmer box and other collection points which will reduce the load on the filter. That in turn will reduce the need to backwash.

Case Study – Penrith City Council reduces Swimming Pools

Penrith City Council is one of the largest councils in Sydney. It owns and operates two Olympic sized swimming pools Ripples Leisure Centre and Penrith Swim Centre. The pools have a total volume including of balance tanks of over 1800 m³ (475.6 thousand US gal.).

An audit conducted by Sydney Water's Every Drop Counts Business Program identified that of all water use in Ripples Leisure Centre, showers accounted for 41.1% and in Penrith Swim Centre, showers accounted for 23.1%. It also identified that make up for backwashing accounted for 28.4% of all water use in Ripples Leisure Centre and 24.1% in Penrith Swim Centre. Reusing backwash water could save Penrith City Council 246 m³ per week.

After the audit the Council put in place the following measures:

- Installed pressure indicator to reduce length of backwash
- Conducting a regular maintenance program to detect and repair leaks
- Monitoring of water meter on a weekly basis to detect unusual water usage
- Calculating L/patron ratio on a monthly basis and comparing with industry benchmarks.
- Monitoring air temperature and humidity for indoor pool.
- Replaced high flow showerheads with low flow showerheads
- Installed flow control in basins.

These measures resulted in reducing water use by 56 m³ /day.

Adapted from: Sydney Water. *The Conserver*. Issue 3. 2003.

12.2.4 Minimising Evaporation

For optimum comfort swimming pools are heated to maintain a temperature of 25° C–40° C (77° F–104° F) [3] depending on the type of pool and activity. Air temperatures are maintained at 24° C–29° C (75° F–84° F) at a relative humidity of 50–60% (for indoor pools). Table 12.4 gives the recommended temperatures for pools. Pool temperatures are also a function of the age and gender of the users. Males and the young require less heat. By selecting the optimum temperature for the activity and the age group greater comfort can be provided to customers.

Given these temperatures and the high surface area, a swimming pool is a storage area for energy with heat gains and heat losses occurring continuously. Evaporation accounts for 70% of energy losses in both outdoor and indoor swimming pools as shown in Figure 12.2 [4]. Evaporation reduces the pool temperature. Whilst it takes only 4.2 kJ to heat one litre of water by 1°C it requires a further 2260 kJ to evaporate each kilogram of water. Besides temperature pools also lose water and chemicals. Figure 12.2 shows the major areas from where energy losses occur in indoor and outdoor pools.

Evaporation is increased by:

- high pool water temperatures
- high air temperatures
- low relative humidity
- high wind speed at the pool surface.

In indoor pools, higher the evaporation of water , the greater the concentration of chemicals, heat and humidity in the air. The pool ventilation system is the only means of removing the contaminants and this then increases energy costs. Increasing the pool temperatures by 1°C increases energy costs by 10–15% [5], since increased evaporation leads to increased ventilation rates to maintain the relative humidity at 50–70%.

The rate of evaporation in kg/h m^2 can be estimated for a swimming pool in normal activity, integrating splashing due to the bathes on the accesses of a limited zone (Smith, et al., 1993) (ASHRAE, 1995), according to the following formula [6]:

$$w_p = A(p_w - p_z)(0.089 + 0.0782\,V)/Y \tag{12.1}$$

Table 12.4 Recommended pool water temperatures

Pool type and use	Recommended temperature range (°C)
Spa	32–34
Hydrotherapy	27–30
General swimming	24–27
Competition swimming	20–23

Courtesy of Australian Greenhouse Office. Swimming Pools and Leisure Centres E – 17.

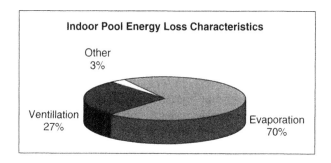

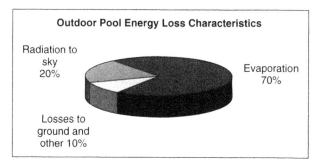

Figure 12.2   Energy loss from indoor and outdoor pools [4]

Courtesy of US Department of Energy. Energy Efficiency and Renewable Energy.

where

A area of pool surface, m^2

p_w saturation vapor pressure taken at surface water temperature, kPa

p_a saturation pressure at room air dew point, kPa

V air velocity over water surface, m/s

Y latent heat required to change water to vapor at surface water temperature, kJ/kg.

In outdoor pools, evaporation is highest when the difference between the water and the ambient temperature is at a maximum – at night. The drier the air (lower the humidity) the greater the evaporation rate.

Higher the air movement above the pool surface increases the evaporation because it removes the warm moist air just above the pool surface. A similar effect is produced by increased ventilation.

For an example, in outdoor pools in Sydney during the warm summer months an average of 6.4 mm of water m^2/day is lost to evaporation [7].

Evaporation can be minimised by installing pool covers. A well-fitting pool cover provides an impermeable layer that virtually eliminates evaporation as well as reducing heat loss through convection and conduction.

Pool covers are made of special materials, such as UV-stabilized polyethylene, polypropylene or vinyl. They can be transparent or opaque. Covers can even be light- or dark-coloured.

Bubble covers are a favourite choice. These have a thickness of 200–500 mm and a life span of 5–10 years. Vinyl covers and insulated vinyl covers have a longer life than bubble covers.

Pool covers are available as manual, semi-automatic and automatic. Semi-automatic covers use a motor-driven reel system. They use electrical power to roll and unroll the cover, but human intervention is required to pull and guide the cover. Semi-automatic covers can be built into the pool deck surrounding the pool or can use reels on carts.

Automatic pool covers are the most convenient of the pool covers for commercial establishments. Automatic covers have permanently mounted reels that automatically cover and uncover the pool at the push of a button. They are the most expensive option, but they are also the most convenient.

The benefits of using pool covers are

- 95–97% of the water loss from evaporation leading to a saving in make-up water of 30–50% [4].
- Reduces energy consumption by 50–70% [4].
- Reduce the pool's chemical consumption by 35–60% [4].
- In outdoor pools, reduces the cleaning time by keeping dirt and other debris out of the pool.

Points to consider when purchasing a pool cover are

- Storage place for the pool cover. The location can increase installation costs.
- Not all pools will be suitable for pool covers. Leisure pools that have complicated contours will be difficult to cover.
- The number of pool attendants present at night will dictate whether a manual or an automatic system is suitable. For example, if only one or two pool attendants are present, then an automatic system may be the most suitable. This is an important decision since if the pool cover is not used then no savings can be expected.
- Increased savings can be expected if the ventilation fans are single-speed rather than variable-speed drives.
- Safety aspects need to be investigated. Can the pool cover support the weight of a person in case someone accidentally fell into the pool?

Case Study [8]

Eastern Leisure Centre in Cardiff, UK, has shown that by installing a semiautomatic pool covers the Centre was able to reduce its energy consumption by 22% and 15% of its energy cost. The staff covers the pool every night for 8 hrs and switch off the heating and ventilation systems.

This has resulted in a saving of £80 000 over a 10-year period. The payback for the original pool cover was 1.6 years.

Another observation was that over this period the Centre has noticed a marked reduction in the deterioration of the building fabric.

References

[1] Energy Efficiency Best Practice Programme. *Good Practice Guide 219. Energy Efficiency in Swimming Pools – for Centre Managers and Operators*. Hartwell Oxfordshire. September 1997.

[2] Sydney Water. *Saving Water in Community Swimming Pools or Leisure Centres*. March 2005.

[3] Energy Efficiency Best Practice Programme. – CTV 006. *Sports and Leisure Introducing Energy Saving Opportunities for Business*. www.carbontrust.co.uk/energy. March 1996.

[4] US Department of Energy. *A Consumer's Guide to Energy Efficiency and Renewable Energy: Your Home – Swimming Pool Covers*. www.eere.energy.gov/consumer/your_home.

[5] Energy Efficiency Best Practice Programme. *Good Practice Guide 228 – Water Related Energy Savings – A Guide for Owners and managers of sports and leisure centres*. Hartwell Oxfordshire. March 1996.

[6] Lund J.W. *Design Considerations for Pools and Spas*. Geo-Heat Centre Bulletin. Oregon Institute of Technology. Klamath Falls, Oregon. September 2000.

[7] Dawson P. *How Much Water Does a Pool Use after Filling*. Sealed Air Corporation Australia.

[8] Energy Efficiency Best Practice Programme. *Good Practice Case Study 76 – Energy Efficiency in Sports and Recreation Buildings: Swimming pool covers*. Hartwell Oxfordshire. September 1994.

Case Study: UK Soft Drinks Industry

In the United Kingdom over 25 billion litres of water is used to produce 10 billion litres of soft drinks that are consumed each year in the United Kingdom. Production, water and wastewater discharged, how the water is used by each category, and specific water consumption and wastewater discharged breakdown is given below.

	Production (%)	Water use (%)	Wastewater discharge (%)
Carbonates or dilutables	84	57	42
Carbonates/fruit juices	8	32	48
Bottled waters	5	5	4
Fruit juices	3	6	6

	Carbonates or dilutables category (%)	Carbonates/ fruit juices category (%)	Bottled water (%)	Fruit juices (%)
In product	78	23	30	27
Equipment preparation	3	7	67	51
Boiler water	4	7		11
Pasteurisers	6	4		4
Cooling water		2		4
Floor washing	1	2		3
Rinsing containers	4	20	2	
Domestic use	3	2	1	
Bottle washing		33		
Other uses	1			
Total	100	100	100	100

Category	Specific water consumption m³ water/m³ product	Specific wastewater discharged m³ water/m³ product
Carbonates or dilutables	2.3	1.4
Carbonates/fruit juices	6.1	3.6
Bottled waters	1.6	0.8
Fruit juices	3.5	1.5

Adapted from: Envrionmental Technology Best Practice Programme: EG 126 – Water Use in the Soft Drinks Industry. June 1998.

Figure 13.1 shows the water usage intensity of the Australian F&B industry over a 3-year period [7].

Raw water quality is also an important factor for this industry especially in the beverage sector. Many facilities have stringent water quality criteria which sometimes exceed the local water authority's water quality standards.

Additionally, the food processing industries also discharge effluent of high organic strength to public sewers. Consequently, the industry is under increasing pressure to become water efficient as well as pretreat effluent to meet regulatory standards. For an example, in Europe, the European Union legislation governing the discharge of industrial effluents require that by 2007, industries comply with directives such as the Integrated Pollution Prevention and Control (IPPC), Water Framework [8]. The IPPC requires industries to install equipment that satisfies the requirements under the Best Available Technology (BAT) directive. Similar concerns are expressed in the annual survey of members of the Australian Food and Grocery Council as shown in Figure 13.2 [7]. After packaging, 47% of respondents voted water as the next highest priority.

13.2.2   Energy Usage

Energy is a vital input in the food processing sector. For instance, breweries in the United States spend in excess of US $200 million on energy. Electricity is used for pumping and conveying fluids, refrigeration and cooling. Steam is used for heating, drying and evaporating products and for generating hot water for washing. For instance, in the dairy industry, approximately 80% of the plants needs are met by the combustion of fossil fuel to generate hot water and steam. Electricity costs account for over half the energy costs. F&B Plants rarely cogenerate their own electricity with steam. Steam is generated for process use with the exception of breweries. Breweries and dairies can cogenerate electricity with extraction steam turbines. There are also variations in energy usage within the same industry sector. Energy use will depend on the product mix. Processes which depend on drying and evaporation are more energy intensive. For instance, in the milk industry, production of whey is more energy intensive than production of market milk. Similarly energy usage of draught beer is less than for canned beer. Other factors that impact on energy use are

- The size of the production plant. Larger plants tend to use less energy per unit of product.
- Age of the plant.
- Level of automation.
- Operation and maintenance practices.

Given the large reliance on water to generate steam, hot water, washing and sanitation and electricity for refrigeration, energy reduction measures go hand

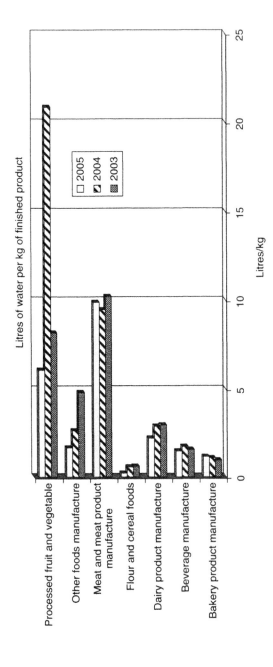

Figure 13.1   Water usage in Litres per kg [7]

Adapted from the Australian Food and Grocery Council's 2005 Environment Report.

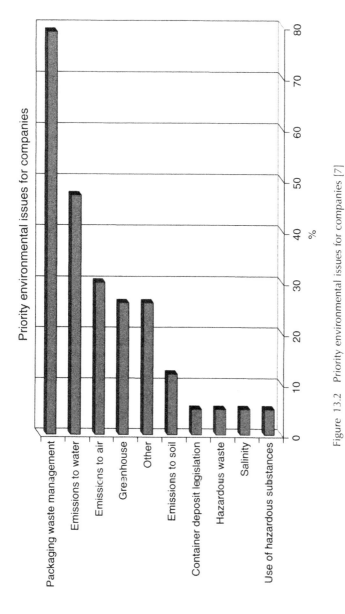

Figure 13.2   Priority environmental issues for companies [7]

Adapted from the Australian Food and Grocery Council's 2005 Environment Report.

in hand with water conservation strategies. Strategies include capturing waste heat from ammonia refrigeration systems to produce hot water, condensate capture and optimisation of the steam and refrigeration systems.

To become water efficient it is necessary to undertake the steps given in Chapter 3.

These are

- understanding the process and determining where water is used
- benchmark water usage and compare it against best practice
- identify water-saving measures.

13.3   Understanding the Process and Where Water is Used

Figure 13.3 gives a typical breakdown of average water usage in the food-processing sector based on 22 audits carried out by Sydney Water's Every Drop Counts Business Program. The average water usage in these facilities was 474 m³/day. The pie chart shows that the majority of the water goes into the processing of the food rather than the product itself, followed by washdown including clean in place (CIP), product usage, utilities, leakage and lastly amenities.

A flow diagram is a good way to understand how water is used in industry. Figure 13.4 shows the flow diagram of a typical poultry processor. These steps are further described in Table 13.1. Therefore, using poultry processing as an example, a business could target the high water-using areas for water minimisation.

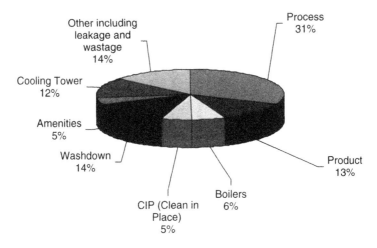

Figure 13.3   Breakdown of water usage in a typical food-manufacturing facility

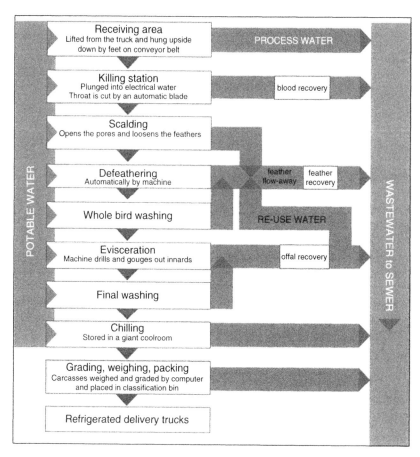

Figure 13.4   Flow diagram of water usage in a typical poultry-processing plant
Courtesy of Sydney Water *The Conserver* 2006. Issue 10.

13.4   Benchmarking Water Usage and Comparing it Against Best Practice

Benchmarking water usage can be done either using kg or litres (lbs or gallons) of raw product or as finished product or as whole units such as L/ bird. There is no accepted convention.

It is important to remember that when comparing against other plants, there will be discrepancies since it depends on the following factors:

i)   Process type. For example, in the poultry industry chilling can be done by using air chilling or immersion chilling. The water usage for these two processors will be dramatically different.

ii)   Animal size and type.

iii)   Slaughter technique

Table 13.1 Description of poultry processing

Step	Description
1. Receiving Area	Chickens removed from crates/cages and placed in shackles (hung upside down). Crates are washed using automatic crate washers, semi-automatic or manually.
2. Stunning	Birds are electrically stunned by dropping them into a water bath, although other methods are available such as gassing (for turkeys).
3. Slaughtering	Birds are slaughtered by cutting the neck and bleeding out (typically 2–3 minutes). The blood is collected and removed.
4. Scalding	Carcasses are immersed in a hot water tank at temperatures of 50–65° C (122–149° F) to loosen the feathers to facilitate plucking. Temperature of the scald tank is critical and varies depending on poultry species and production methods (i.e. needs to be high enough to loosen feathers but not too high as to damage the carcass). USDA requirement is 0.9 L/bird (0.25 US gal./bird) [9].
5. Defeathering	Feathers are removed from the carcass – using equipment comprising a bank of counter-rotating steel discs (automated production line) or rotating steel drums (manual production) with mounted rubber fingers.
6. Washing	Water is constantly sprayed to flush away removed feathers. Remaining feathers are removed by hand.
7. Evisceration	Evisceration involves cutting around the vent and insertion of a spoon-shaped device to remove the viscera. Can be done either mechanically or by hand, but care must be taken to ensure the viscera is not damaged or ruptured as this can lead to significant contamination of the carcass.
8. Washing step	Eviscerated carcass is washed internally and externally.
9. Chilling	To minimise microbial proliferation from *Salmonella and Camhylobacter,* the carcass is chilled immediately to a temperature of 4° C (40° F). Removal of carcass heat using air-chilling, water immersion (spin chilling) or spray chilling. Water immersion chilling is the most common method practiced in the US and Australia, with the carcass placed in counter-current flow of chlorinated (50–70 ppm total available chlorine, 0.4–4.0 ppm free available chlorine) cold water ($\sim 0°$ C). In Europe air chilling is more popular. Water flow rate for immersion chilling differ from country to country. USDA requirement is for 1.9 L/bird (0.5 US gal./bird) [9] to 5 L/bird. Water discharged contains parts of flesh, grease and blood.
10. Grading Weighing and Packing	Birds are replaced on overhead conveyors to allow excess moisture to drain. No fresh water is used. Giblets are put back and weighed.

    iv) Degree of automation
    v) Cleanup and housekeeping procedures
    vi) Conveyance means.

The benchmark need not be against an industry standard but can be a year-on-year comparison for the whole site, process or product. This eliminates differences in production processes. Table 13.2 gives typical and best practice figures for selected industries.

Table 13.2 Typical and best Practice figures for selected industries

Industry	Typical water use	Best practice water consumption
Chicken	13–37.8 L/bird [9]	8–15 L/bird [10]
Turkey	41–87.0 L/bird [9]	40–60 L/bird [10]
Dairy	1.3–2.5 L/L of raw milk	0.8–1 L water/L raw milk [11]
Ice cream	3.6–10.3 L/kg	
Red meat abattoir	6–15 m³/ton hot standard carcass weight (HSCW) [12]	8 m³/ton HSCW [12]
Beer	5–7 L/L of Beer [13, 14] For bottled beer – 10 L/L of beer	2–3.5 L/L of Beer [13]
Soft drinks – Cola	1.5 L–3.9 L/FBL [15]	1.30 L/L per finished beverage (FBL) [16]

Case Study: Coca-Cola Amatil Northmead Plant Australia – Achieving Best Practice

The Coca-Cola Company is the world's largest beverage manufacturer with production plants in 741 locations. The Global Water Initiative was launched in 2004 to reduce water usage and since then has made steady progress. Coca-Cola Amatil's Northmead plant since 2003 has saved over 230 m³/day of potable water. Through a partnership with Sydney Water's *Every Drop Counts Business Program* in 2004 they reduced their benchmark to 1.37 L/FBL.

Coca-Cola Amatil's water usage benchmark is compared to other Coca Cola plants in the table below.

Country	Water usage in 2005 L/FBL (finished beverage litre)
Coca-Cola Amatil Australia	1.58 (actual)
Coca-Cola UK	1.50 (target)
Coca-Cola Enterprises Inc, USA	1.86 (actual)
Coca-Cola Company (average for all worldwide 741 beverage production locations)	2.60 (actual)
Coca-Cola India	3.90 (2004 usage)

Sources:
Coca-Cola UK 2005 Environment Report
Coca-Cola Company 2005 Environment Report
Coca-Cola Enterprises Inc. 2005 Environment Report
Coca-Cola Amatil website –www.ccamatil.com/AusWaterRedTargets.asp

> **Note to the reader:**
> However in making these comparisons it needs to be borne in mind
> that the product mix and age of the plant may make direct comparisons
> difficult and therefore is only given as a guide and not to be taken that
> every plant can achieve the same efficiencies
>
> *Courtesy of Coca-Cola Amatil Australia.*

13.5 Water Minimisation Measures in the Food-Processing Industry

Tables 13.1 and 13.2 show that most of the water usage is for cleaning operations to maintain sanitary conditions in order to comply with health, integrity
of the product and food safety regulations. Therefore water-use efficiency
and reuse measures need to be considered in the light of not compromising food safety whilst minimising water usage. In assessing opportunities,
consider the water-minimisation hierarchy of avoid, reduce, reuse and recycle as shown in Figure 13.5. However, recycle is not a practical option for
food-processing plants.

13.5.1 Avoiding Water Usage

The objective of preventing food or other waste from becoming water-borne
is to keep *dry waste dry and wet waste wet*. Approaches currently used
include the following.

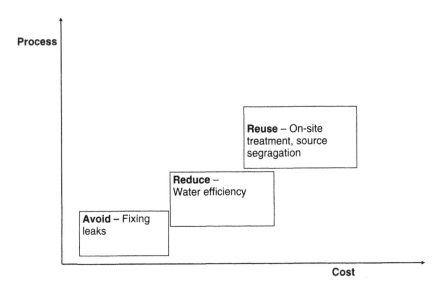

Figure 13.5 Water-minimisation hierarchy.

- Install sub-meters to detect leaks. Connect these to an electronic monitoring system so that excessive water usage from leaks, tank overflows and faulty valves can be detected and acted upon immediately.
- Use brooms, brushes, vacuum cleaners and squeegees before floor rinsing.
- Dry-cleaning vessels before rinsing to remove solids for recovery or disposal.
- Change of production procedures to minimise product or by-product wastage such as installing spill collection trays to collect solids at appropriate places in the production line.
- Install solenoid valves to link sprays to machines so that the sprays operate only when the machines are running.
- Link operation of vacuum pumps to usage through pneumatic timer switches.
- Water-based conveyor systems can be replaced by mechanical systems. "Pigging" of lines rather than using water to clean lines. When multiple products are made in the process vessels it is necessary to clean the lines before sending another product. By sending a silicon bullet, it pushes the extra product down the line and into the processing equipment negating the need to use high pressure jets.
- Use air-cooled chillers to replace wet-cooling towers and evaporative condensers if practical.
- Air thawing or microwave technology rather than water thawing if practical.
- Air chilling rather than water chilling.
- Using synthetic lubricants rather than water-based lubricants.
- Repair leaks.
- Dry peeling rather than wet peeling
- Air rinsing of bottles and cans rather than with water.
- Replacing liquid ring vacuum pumps with dry vacuum pumps (see Section 13.4.5)
- Reduce the size of nozzles if practical.

The added advantage from these measures is that the wastewater loads are reduced. Given that food-processing plants have biological loading, frequently this becomes the driving force for such capital investment.

Case Study: Air rinsing of bottles and cans at Diageo, Huntingwood, Sydney, Australia.

Diageo is the world's largest manufacturer of alcoholic beverages. In the past three years Diageo's Huntingwood site has trebled its production output while reducing its water product ratio by 30%.

One of those measures was to change over to air rinsing of bottling and can lines. By December 2004, three of the four lines have been converted to air rinsing. Saving $36\,000\,m^3$/yr of water. A 20% reduction in water usage.

"We regard water as a key performance indicator," said plant manager Chris King. "We have reduced our use of water from 2.1 L per product in 2001 to less than 1.4 L per product in 2004."

Adapted from: Sydney Water. *Raising a glass for Water Conservation. The Conserver*, Issue 6. December 2004.

Case Study: Hans Continental Foods, Sydney – Hot air thawing

A water audit showed that one-third of all water is used for thawing of frozen meat products with continuously running water. The company commissioned a new process – controlled atmospheric thawing which passes hot air in a temperature- and humidity-controlled environment.

The benefits were immediate. The plant cut its water use by more than 30% and is now saving more than 130 000 L/day and at least $100 000 annually.

The new process also delivers better results with a higher level of retained proteins. Plus it adds extra capacity to the plant, improves the quality of the wastewater and reduces the cost of wastewater treatment.

Adapted from: Sydney Water. *The Conserver*. Issue 7. April 2005.

13.5.2 Reducing Water Usage – Spray Nozzles

Spray nozzles are used widely for cleaning and a variety of other applications within the food-processing industry. And yet they are one of the most overlooked items. It is reported that a reduction of 20% in water usage can be achieved by improving the spraying systems. The benefits of using spray nozzles are

- efficient and effective cleaning
- reduced water consumption
- reduced operating costs
- reduced cleaning time
- ability to restrict sprays to specific flow rates.

To achieve these benefits spray nozzle selection needs to be based on

- flow rate – volume of fluid sprayed at a given pressure
- spray pattern – the dimensions and uniformity of coverage

- pressure and pressure drop
- droplet size – particle size
- the material to be cleaned
- spray impact and spray velocity
- matching cleaning fluid characteristics with nozzle material.

Failure to consider the above factors can result in the following:

- High flow rates increases moisture in the air, creating a hospitable environment for micro-organisms to persist and even proliferate.
- Incorrect selection and incorrect use such as high pressure flows may disperse micro-organisms to the air (aerosols).
- Using the wrong spray pattern. For quick cooling a hollow cone with large droplets may be better suited than other patterns.
- Misting and overspray results when the droplet size is too small.
- Improper placement of spray nozzles results in poor coverage of target area.
- Abrasion from particles in the cleaning fluid can increase nozzle wear and tear.
- Aluminium and brass nozzles are cheaper than stainless steel and hardened stainless steel nozzles. However, the latter are more abrasion resistant than aluminium or brass nozzles and therefore will provide better performance over time easily justifying the added expense. Table 13.3 shows the approximate abrasion resistance ratios of typical spray nozzle materials.

(i) Use the recommended spray for the application – The proper spray pattern is required to achieve the cleaning objectives. A piece of

Table 13.3 Approximate abrasion resistance ratios of typical spray nozzle materials

Material	Abrasion resistance ratios
Aluminium	1
Brass	1
Steel/Iron	1.5–2
MONEL®	2–3
Stainless steel	4–6
HASTELLOY®	4–6
Hardened Stainless Steel	10–15
STELLITE®	10–15
Silicon Carbide (Nitride bonded)	90–130
Ceramics	90–200
Carbides	180–250

Courtesy of Spraying Systems Pty Ltd. – Spray Nozzle Maintenance Handbook. 1992.

metal pipe with drilled holes may lead to excessive water usage and increase cleaning times. Excessive water and hot water usage not only increases the dissolution of meat, fat and other valuable food ingredients but also increases hydraulic and pollutant loadings of the wastewater treatment plant. This in turn increases trade waste charges and compliance fines. The case study given below illustrates how Coca-Cola Amatil in Sydney NSW was able to save $78 500 yr.

Case Study: Benefit to Coca-Cola Amatil

Coca-Cola Amatil Northmead site in Sydney, Australia, has been actively seeking ways to reduce water consumption. One such opportunity was discovered at the bottling rinsing area. Before filling bottles with product, rinsing with treated town water is required. The rinse water is then sent to waste treatment and finally to sewer.

It was recognised that the rinser nozzles were not designed with water efficiency in mind. Coca-Cola Amatil sourced a spray nozzle that delivered a far more efficient spray, providing excellent coverage, yet minimising water usage.

The new nozzles reduced water usage on two lines by 176 L/min – a reduction of 46%! The annual projected savings are 37 400 m^3 and $78 500. Plans are now in place to retrofit spray nozzles in their Smithfield plant.

Source: Sydney Water Fact Sheet. *Spray Nozzles*

(ii) Maintenance of Spray nozzles – Spray nozzles become clogged, corrode or get damaged due to heat. The end result is spray pattern that is not ideal either wasting water or not performing to design. When sprays become clogged then drills are commonly used to clean them. This increases the nozzle size. Doubling the diameter quadruples the water flow rate. Therefore it is important to inspect them regularly and replace when necessary.

Nozzle maintenance check list should include

- measuring flow rate
- checking spray pattern
- checking spray pressure.

Figure 13.6 shows a comparison between clogged, corroded and a new nozzle.

A simple example can indicate just how much poor maintenance could be costing a company.

Figure 13.6   A Comparison between a good flat spray pattern (left) and a spray from a clogged nozzle and photos of a corroded and clogged flat spray nozzles.

Courtesy of Spraying Systems Pty Ltd.

(iii) Automation of sanitising regime – In food-processing plants it is a common practice for sanitising to be done at night using portable sprayers. Manual application results in inconsistent spray applications, labour time and wastage of chemicals if over sprayed, and if under sprayed sanitation is compromised. In these applications proper spray application is dependent on the operator. An investment in an automated fogging system with a dedicated spray controller and air atomising nozzles will improve the situation.

Worked example: Cost of worn spray nozzles

Number of sprays	30 nozzles
Operating time	16 hrs/day
Number of days operated/yr	300
Typical nozzle flow	10 L/min
Typical wastage with worn nozzle	15%
Increased water wastage	12 960 m^3/yr
Cost of additional water	$31 104/yr*

*At $2.40/m^3 of water and wastewater. Additional power and chemical costs for effluent treatment have not been included.

13.5.3   Reducing Water Usage – Washing

Bottle washing is an area where there can be a high demand for water. The common causes are excessive flow rates; lack of solenoids in valves and spray nozzles having too high flow rates (read the Coca-Cola case study). Install timers on lines so that during non-operating hours water is not wasted. Investigate the possibility of air rinsing rather than water rinsing. Only final rinse needs to be with fresh water. For pre-rinse reuse the final rinse to capture water and heat.

For crate washing consider installing an automatic crate washer.

Case Study: National Foods installs automatic crate washer and a unique tanker washer

Washing crates is part and parcel of National Foods, Australia's largest dairy's daily routine. At it's Penrith Plant in Sydney, a large portion of the plant's total water usage is in the washing of over 34 000 crates per day. Around 60 m³/day went straight to the drains but is now filtered and fully reused, delivering savings of over A\$ 100 000 per annum.

At the milk intake point, savings are being made in both washing tankers internally and externally. A dual-washing system that used over 1000 L/truck has been replaced with a purpose built system that uses just 70 L and is more effective and twice as fast. The subsequent truck washing is now controlled by a time limited token system.

This and other water reduction measures has helped National Foods to better their benchmark for market milk beyond international best practice water usage target of 0.9–1.0 L/L.

Source: Sydney Water. *National Foods Timely Management Delivers Big Cost Reductions*. The Conserver. Issue 4. 2004.

For cleaning of floors and equipment install trigger-operated spray guns to hoses. A common model of a trigger-operated spray gun will deliver 20 L/min compared to a standard hose flow rate of 30 L/min. The problem associated with spray guns like the type shown in Figure 13.7 is theft. These are more expensive than domestic trigger guns and therefore are an easy target for theft. Some employees may resist using the guns. Education and training will help in these instances. One company solved the problem by giving each operator a domestic trigger gun before installing the industrial variants.

13.5.4   Reducing Water Usage – Clean-in-Place

Clean-in-place (CIP) technology offers significant advantages to food-processing and other manufacturing plants to efficiently and economically clean process equipment, tanks and piping, improved hygiene and product

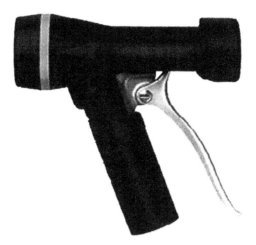

Figure 13.7   A Water-efficient spray gun

Courtesy of Spraying Systems Pty Ltd.

quality. The ability to do so without dismantling plant equipment significantly reduces cleaning-time efficiency and increases operator safety through reduced chemical handling allowing the use of higher strength detergents and hotter temperatures.

The cleaning solutions are generally distributed to the CIP circuits from a central CIP station consisting of several tanks for storing of the cleaning solutions. The solutions are heated by steam and their concentration is constantly monitored and adjusted. The cleaning programme differs according to the equipment to be cleaned, but the main steps are

- Pre-rinse cycle – Soiled equipment is cleaned with warm water for 3–10 minutes to remove loose solids.
- Cleaning cycle – Removal of residual solids from equipment surfaces by optimising the four factors mentioned above. An alkaline detergent (typically 0.5–1.5% sodium hydroxide) at 75° C is circulated for about 10 minutes or longer.
- Post rinse – Rinsing all surfaces with cold to hot water depending on the temperature of the cleaning cycle to remove residual chemical solutions and contaminants.
- Acid rinse – A mild acid rinse to neutralise any alkaline residues. Hard to remove proteins may require other additives (peptising) and wetting agents to aid removal.
- Disinfection – Normally carried out immediately before the production line is to be used again. Hot water at 90°–95° C or chemical disinfectants are used. Chemical disinfectants can be sodium hypochlorite, peracetic acid (PAA) or quaternary ammonium compounds.

- Cooling – Cooling with water to rinse out disinfectant residuals (to eliminate the chemical contamination of food and to minimise corrosion
  to piping and equipment) or cooling of equipment.

The designs vary from single-pass to recirculating systems. The single-pass
systems whilst inexpensive and have a smaller footprint uses more water
chemicals and heat. Figure 13.8 shows a schematic of a single use CIP
system.

The operation of a CIP system requires the control of several conditions.
The five factors that need to be considered when designing a CIP system are

- Time – The longer a cleaning solution remains in contact with the
  equipment surface, the greater the amount of soil that is removed.
- Chemical concentration – These vary depending on the type of chemical used, the type of contaminant and the equipment to be cleaned.
- Temperature – Soiling is affected by varying degrees by temperature. In
  the presence of a cleaning solution most contaminants become more
  soluble as the temperature is increased. The temperature can be as
  high as 95° C.
- Fluid velocity – This aids in removal and typically reduces time, temperature, and concentration requirements. Fluid velocities in process
  piping are approximately 1.5 m/s (5 ft/s) or higher. In tanks the flow
  rate may be around 40–600 L/min.
- Spray design – Spray designs may range from small static nozzles to
  rotational (360°).

These requirements are dictated from sanitising requirements imposed
by health authorities. Sanitising of food product contact surfaces means
reducing bacterial counts by 99.999% (5 log) reduction. And sanitising non-product contact surfaces require a 99.9% (3 log) reduction of
contamination.

Given the importance of CIP, the opportunities to save water in CIP
systems are

(i) Internal recycling
     Water, chemical and heating costs can be reduced by
     - Collecting rinse water for reuse and recirculating chemical solu-
       tions back to a storage tank. Periodic inspection and cleaning of
       rinse water collection tanks to prevent the build-up of unwanted
       deposits is required.
     - Collecting hot rinse water maintains equipment temperature,
       allowing caustic to be introduced more quickly.
     - Conductivity sensors rather than timers to be used to divert caustic
       detergent to the storage tank instead of the drain.
     - Using membrane filtration to clean contaminated caustic avoiding
       the need to dispose of the caustic to the drain.

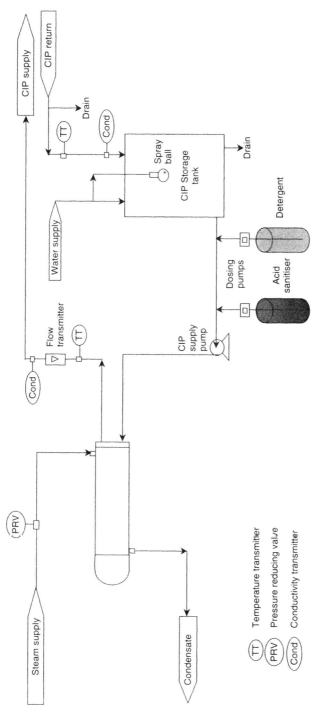

Figure 13.8   Schematic of a single use CIP skid system

(ii) Optimising CIP programmes
Typically 20% water savings can be achieved without compromising cleanliness standards by
- By customising the CIP programme for the size of the plant and type of soiling. Often pre-rinse and post-rinse times are not appropriate for the plant and size of equipment. Consult a reputable CIP detergent supplier to carry out an audit of the CIP system.
- Using software to monitor CIP systems. Software systems provide real-time monitoring and data storage, charting of trends to analyse the performance of the system.
- Improving product scheduling so that similar products are processed sequentially, reducing cleaning requirements.

(iii) Using spray designs that are fit for purpose
Water use can be reduced by using spray nozzles that are fit-for-purpose with regard to spray pattern, temperature and chemical corrosion of materials.
- Self-powered rotating nozzles have a superior performance over fixed spray nozzles as well as using less cleaning chemicals. Figure 13.9 shows a photo of a rotating spray nozzle.

(iv) Removing product and gross soiling before cleaning
Removal of residual product before cleaning with CIP system will save product going to the sewer, water, chemicals, reduce pre-treatment chemical costs as well as trade waste charges. Pigging and air-blowing techniques are commonly used.

(v) Ensuring equipment is correctly designed for CIP cleaning
When a CIP system is connected to the existing pipework or when designing a new processing plant, ensure to minimise 'dead legs', crevices or pockets that cannot be reached by the cleaning solutions. Detergents and sanitisers are not able to reach these areas and thus allow micro-organisms to multiply. Figure 13.10 shows an illustration of incorrect and correct piping design.

13.5.5   *Reducing Water Usage – Liquid Ring Vacuum Pumps*

Liquid Ring Vacuum Pumps (LRVPs) are used extensively in the food industry for evacuation and evaporative cooling. Evacuation is used to empty the vessel before filling with food product. Evaporative cooling is used to cool the food product to minimise spoiling of food. Typical applications of LRVPs include concentrating fruit juice, adjusting the boiling point (temperature) of water during cooking and flash cooling after a food item has been cooked. Common vacuum technologies are

- steam jet ejectors
- barometric condensers

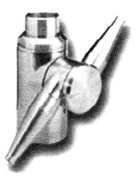

Figure 13.9   A photo of a rotating spray nozzle

Courtesy of Spraying Systems Pty Ltd.

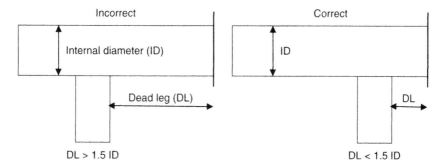

Figure 13.10   Illustration of incorrect and correct piping design

- liquid ring vacuum pumps
- dry vacuum pumps.

Steam jet ejectors use steam through a nozzle which discharges a velocity jet across a chamber that is connected to the equipment to be evacuated

Barometric condensers create a vacuum by condensing the vapour through contact with cooling water.

In an LRVP as the impeller of the vacuum pump rotates, it throws water by centrifugal force to form a liquid ring concentric with the periphery of the casing which does the work of compression. The eccentric mounting of the impeller with respect to the casing results in increased spacing between the impeller blades at the inlet port and the decreased spacing towards the outlet port. As the vapour (gas) enters the inlet port, it is trapped between the impeller blades and the liquid ring. Then the impeller rotates, the liquid ring compresses the vapour or gas and forces it out the outlet port. Figure 13.11 shows the principle of operation of a liquid ring vacuum pump.

The working principle of the LRVP requires a continuous supply of sealing liquid and in many cases this is water. The sealing water maintains the seal,

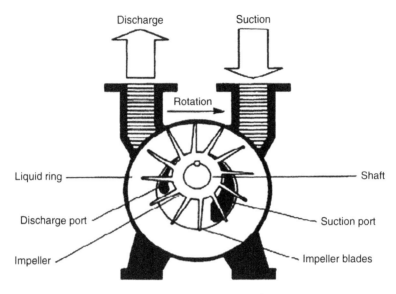

Figure 13.11   Principle of operation of a liquid ring vacuum pump

absorbs the heat of compression, friction and condensation. The clearances are larger in LRVPs and this feature allows them to handle streams containing particulate matter. The temperature of the seal water is critical to the pump's vacuum-generating performance. Figure 13.12 shows that if the temperature of the seal water increases from 15° C to 20° C at a suction pressure of 40 torr, it will reduce pump capacity by 20%.

To maintain constant temperature, the liquid is often used only once before being discharged to the sewer. Depending on the operating principle, it can waste a lot of water. The three ways that seal water can be connected are

1. once through
2. partial recovery
3. closed loop.

In once-through water systems the water is separated from the gas and discharged to the drain. In the partial recovery about 50% of the liquid is recirculated and 50% is sent to the drain. In the closed-loop system the sealing liquid is cooled (through a heat exchanger using cooling water) and then recirculated.
Water-saving options are

1. Converting from once through to a partial recovery or closed-loop system.
2. Ensuring that the vacuum pump is not oversized for the application. To overcome increases in process temperature and a consequent loss in vacuum, LRVPs are frequently oversized.

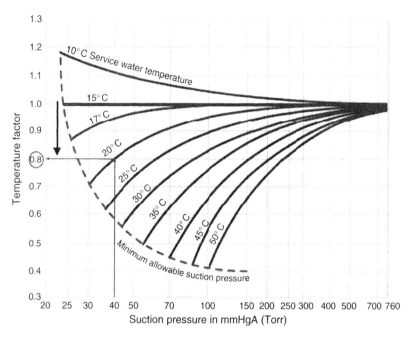

Figure 13.12 Chart showing the relationship between seal water temperature and pump capacity.

Courtesy of Sydney Water. Liquid Ring Fact Sheet October 2004.

3. Investigate the option of changing over to a dry vacuum pump. Dry vacuum pumps require no sealing water. The three types of dry vacuum pumps are hook and claw, lobe and screw. They run hot since there is no liquid to absorb the compression heat and therefore a water jacket will be required. One of the advantages of dry vacuum pumps is that they can handle corrosive vapours by keeping them as vapours and not allowing the vapours to condense. They are also energy efficient since they move air not water. To compensate for the decrease in efficiency, frequently LRVPs are oversized. On the other hand, a claw-type dry vacuum pump is about the 50% the size of an equal LRVP. On recirculated water systems, energy is also consumed to cool and move water around the plant. Since the dry vacuum pumps do not require water for sealing, they can be fabricated of inexpensive cast iron.

The disadvantages of dry vacuum pumps are that they cannot handle slugs of liquid or particulates; however, in such cases a knockout pot is required upstream of the pump. Polymerisation of product can be a concern because the pumps are running hot (as much as 315° C). The capital cost of dry vacuum pumps are more than LRVPs. However, the lower operating costs during the life of the pump will make up for the higher capital cost. In some industries, dry vacuum pumps may not be acceptable from a perspective of product purity. Their suitability needs to be individually investigated. Figure 13.13 shows a photo and cross section of a screw-type dry vacuum pump.

Figure 13.13　Photo and cross section of a screw-type dry vacuum pump
Courtesy of Busch Australia Pty Ltd.

Case Study: World's first application of dry vacuum pump at 140° C

Snow confectionery, based in Sydney, is one of Australia's largest producers of sweets, candy, toffees and jellies. LRVPs are used to extract moisture out of the sugar syrup to reduce the cooking time. Their 3 LRVPs were consuming 48 m³/day as sealing water. This was approximately half the site's water usage. The water temperature exiting the LRVPs were as high as 55° C. They were interested in reducing their water usage as well as the effluent discharge temperature. The cause of the high water temperature was the seal water discharged from the liquid ring vacuum pumps connected to the high temperature vacuum cookers operating above 140° C. Dry running vacuum pumps were investigated but they rarely operate at liquid temperatures above 80° C. Busch Australia replaced one LRVP with a dry vacuum pump to test their suitability in this challenging application. To prevent sugar jamming the pumps they designed a system to inject oil to the pump. 12 months later the pump is using only 13.5 kW compared to 17 kW the LRVP was using, the effluent discharge temperatures are lower and when all three cookers are connected water usage will decrease by 50% or 7 million litres a year. The reduction will also assist in Snow Confectionery achieving its target of 1.5 L/kg of confectionery produced.

Adapted from: Waste Management and Environment, *Dry times ahead as costs rise.* July 2006 and Sydney Water. *The Conserver* Snow Confectionery to halve its water use. Issue August. 2005.

13.5.6   *Reducing Water Usage – Amenities Blocks*

Food-processing facilities require the maintenance of good hygiene standards and therefore washing of hands before entering the plant is essential. These facilities can be a source of water wastage. Replace these with on-demand systems or paddle-operated systems. Install meters. Toilet blocks can also be retrofitted. Refer to Chapter 10.

13.5.7   *Reducing Water Usage – Evaporative Condensers and Cooling Towers*

Refrigeration, air-conditioning and compressor systems play a major role in a food-processing plant. Reducing the refrigerant temperature in the condenser is important to reduce energy costs. A 1° C drop in temperature will reduce electricity costs by 2–4%. Maintaining clean heat transfer surfaces in the condenser and optimising water treatment are discussed in Chapters 5 and 6.
   Some common strategies to save water are

- Eliminate once-through water systems.
- Ensure that the cooling towers are not overflowing.
- Maintain bleed levels at the optimum levels.
- Minimise drift or splashing.
- Install variable speed fan drives in cooling tower.
- Investigate the potential for air-cooled chillers or better hybrid cooling towers rather than wet cooling towers. Refer to Chapter 6 for more details.
- Optimise compressed air usage. For every 1 kWh of compressed air used 10 kWh of electricity is required and these require cooling water to keep the system cool.

Refer to the case study on Red Lea Chicken, Chapter 6.

13.5.8 Reducing Water Usage – Steam Systems

Steam systems play a large role in food-processing plants. Chapter 7 discusses some of the common strategies to be undertaken.

A common strategy is to recover steam condensate. Often oil-contaminated condensate is sent to the drain. There are ways of removing the oil and reusing the condensate for its energy- and water-saving value. Capturing flash steam and recovering steam from evaporators are other commonly employed strategies.

13.5.9 Reusing Water

There are several options to reuse water in a food-processing facility. These can be divided into those that do not contact food products and those that do contact food products. Non-contact water-reuse applications are numerous and can easily be implemented. When the reuse water comes into contact with food a key issue that must be addressed and assessed is whether this presents any associated increased risk of microbiological contamination of food and the production environment by pathogenic micro-organisms. Most food-processing companies are reluctant to investigate such options given the risk to their brand name from microbial contamination. Microbial pathogens of concern include *Salmonella, Campylobacter, Listeria*, pathogenic strains of *E. coli, Giardia, Cryptosporidium* and viruses.

To alleviate these concerns regulatory standards require that reuse water meet drinking water-quality standards. A validated and verifiable systematic food safety management tool such as a Hazard Analysis Critical Control Point (HACCP) system will be required to satisfy these food safety obligations. The HACCP system consists of the following seven principles (CODEX Alimentarius Commission, 1997):

- Principle # 1 Conduct a Hazard Analysis.
- Principle # 2 Identify the Critical Control Points (CCP)
- Principle # 3 Establish Critical Limits s at each CCP
- Principle # 4 Establish monitoring procedures
- Principle # 5 Establish corrective action to be taken when monitoring indicates that there is a deviation from a Critical Limit.
- Principle # 6 Establish verification procedures
- Principle # 7 Establish record keeping procedures

The advantage of the HACCP approach is that it incorporates food safety control as a management tool rather than relying only on the end-product testing.

For water reuse the CODEX guidelines specify the following:

- Reuse water shall be safe for intended use and not jeopardise the safety of the product through the introduction of chemical, microbiological or physical contaminants in amounts that represent a health risk to the consumer.

- Reuse water should not adversely affect the quality (flavour, colour or texture) of the product.
- Reuse water intended for incorporation into a food product shall at least meet the microbiological and as deemed necessary chemical specifications for potable water.
- Reuse water shall be subjected to ongoing monitoring to ensure its safety and quality.

Food processors considering the use of reuse water that comes into contact with food are well advised to engage the appropriate food regulatory authority early into discussion.

13.5.10   Notes on Water-Reuse Applications

Appropriate technologies are described in Chapter 8. Turbidity in water is often due to the presence of organic matter. Some decontaminating methods are ineffective in the presence of turbidity to reduce micro-organisms since the microbes can embed themselves inside particulate matter. A comparison of microbial decontamination methods and their sensitivity to turbidity is shown in Table 13.4.

Table 13.4 A comparison of water-reuse technologies for microbial decontamination.

	Membrane processes	Heat treatment	UV – radiation	Hypochlorite	Chlorine dioxide	Ozone
Recommended concentration	NA	NA	25–40 mWs/ cm^2	50– 100 mg/L (total chlorine)	> 2 mg/L	< 1 mg/L
Contact time	NA	Sterilisation requires a temperature of 121° C and a holding time of 20 minutes. Pasteurisation requires a temperature of 65° C for 10 minutes or 72° C for 2 minutes.	0.5–5 seconds	10–20 mins	15 min	2–4 mins
Sensitivity to turbidity	Low. Membranes remove turbidity.	None	Highest	High	Medium	High

End-of-pipe treatment is a more expensive than source reduction. The wastewater of food-processing plants have high organic strength (BOD). Treating this water is expensive. High BOD is frequently an indication of lost product. For an example, in a dairy plant (BOD_5 of milk is 104600 mg/L), it has been estimated that **1 kg of BOD is equivalent to 9 kg of milk lost**. Milk product losses typically range from 0.5% in large modern dairy plants to 2.5% in small plants [18]. Wastewater discharged best practice levels are 0.8–1.7 L/L of milk. One food processors adopted the slogan "Lets not wash our profits drown the drain".

References

[1] *Food production given wake-up call over water scarcity* www.meatprocess.com. September 2004.

[2] Consultative Group on International Agricultural Research (CGIAR). A Third of the World Population Faces Water Scarcity Today. Stockholm, Sweden. www.cgiar.org. 21 August 2006.

[3] Confederation of the Food and Drink Industries of the EU (CIAA). *Industry as a Partner for Sustainable Development.* Beacon Press, UK. 2002.

[4] UNEP Working Group for Cleaner Production. *Eco-efficiency for the Queensland Food Processing Industry.* Australian Industry Group, Canberra. 2004.

[5] Food and Drink Federation_*Industry Statistics.* ww.fdf.org.uk/industrystats.aspx.

[6] Kiepper B.H. *Characterisation of Poultry Processing Operations, Wastewater Generation, and Wastewater Treatment using Mail Survey and Nutrient Discharge Monitoring Methods,* University of Georgia, Athens. 2003.

[7] Australian Food and Grocery Council. *Environment Report.* www.afgc.org.au. 2005.

[8] Industry News. *Food & Beverage: quality, cost and legislation key concerns.* Elsevier: Filtration + Separation. March 4. 2006.

[9] North Carolina Department of Environment and Natural Resources. *Water Efficiency Manual for Commercial, Industrial and Institutional Facilities.* August 1998.

[10] Environmental Technology Best Practice Programme. GG – 233 – Reducing Water and Effluent Costs in Poultry Meat Processing. March 2000.

[11] United Nations Environment Programme Division of Technology Industry and Economics. *Cleaner Production Assessment in Dairy Processing.* 2000.

[12] UNEP Working Group for Cleaner Production. Fact Sheet 7: Food Manufacturing Series: Meat processing, Brisbane. 2004.

[13] Fosters Group Limited. *Environmental Aspects and Impacts.* www.fosters.com.au. 2004.

[14] Sabmiller. *Reducing Our Environmental Footprint.* www.sabmiller.com.

[15] Coca Cola Inc. Environmental Report. www.coca-cola.com. 2005.

[16] Coca Cola Amatil. *Water Reduction Targets.* www.ccamatil. com/AusWaterRedTargets.asp.

[17] Sydney Water. *The Conserver.* December Issue. 2004.

[18] Rausch K.D. and Powell G.M. Dairy Processing Methods to Reduce Water and Liquid Waste Load, Cooperative extension Service, Kansas State University, Manhattan, MF – 2071, March 1997.

Chapter 14

Oil Refining

14.1   Introduction

The oil refining industry is the cornerstone of a modern economy. All countries are subject to the vagaries of global oil prices, which have significant influence on local economies. Current global oil refining capacity is in excess of 11 million metric tons per day (82.5 million barrels/day) [1]. Figure 14.1 shows the world's refining capacity, as well as other major refining processes by continent. Nearly 90% of this is located in non-OPEC (Organisation of Petroleum Exporting Countries) countries [2]. The United States is the world's largest oil refiner with 149 of the world's 691 refineries and a crude oil refining capacity of about 2.3 million metric tons/day (16.9 million barrels/day) followed by Russia's – 0.73 million metric tons/day (5.4 million barrels/day), Japan – 0.65 million metric tons/day (4.7 million barrels/day) and China – 0.62 million metric tons/day (4.6 million barrels/day) [2].

There are major challenges facing the refining industry in the 21st century. The industry is unique in that both the processes used to refine petroleum as well as the products generated, are subject to government regulation. Some of these challenges are:

- Addressing the long term decline of easily accessible resources
- Demand for more energy
- Decrease in crude quality (heavier, higher sulphur and metal concentrations more prevalent)
- Climate change and pressure from government and environmental groups to reduce CO_2 emissions and Nitrogen oxides
- Meet new emission standards for diesel and gasoline
- Low refinery margins (until recently) and over capacity
- Competition from newly developing countries (China and India)

The positive aspect of these challenges has been to shut inefficient refineries, grow through consolidation or acquisition of more efficient refineries and invest in new environmentally benign processes. The downside is that some of these processes use more water.

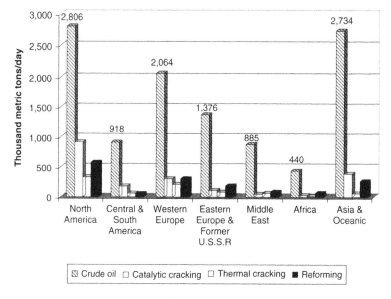

Figure 14.1   World refining capacity in 2004 [1]

(1 Metric ton of crude oil = 7.33 barrels)
Adapted from Energy Information Administration.

For example, in the U.S. between 1981 and 2003, from just over 300 refineries only 149 are now operating [1–3]. In Europe, between 1991 to 2000 only two refineries have been built. While the number of refineries declined, the overall crude processing capacity has increased by between 1 and 2%, due to existing refineries modifying their processes. Figure 14.2 shows the distribution of refineries by continent [1].

The energy industry is dominated by a few large integrated oil companies able to withstand the political and cyclical nature of the business. ExxonMobil is the largest and revenue for the year 2006 was reported at

Number of Refineries

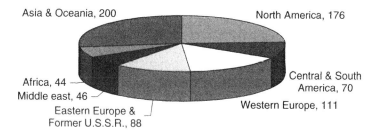

Figure 14.2   Distribution of refineries by continent (2004)

Adapted from Energy Information Administration.

US$ 39 billion which is greater than the combined economies of most developing countries. Other oil majors are Royal Dutch Shell, BP, Chevron and Total.

14.2   Oil Refining Processes

Crude oil is essentially a mixture of hydrocarbons – aliphatic and aromatic hydrocarbons. It also contains many impurities, such as organic sulphur, nitrogen, heavy metals, salt and sediment. Crude oil therefore needs to be washed to remove the salt and sediment and then separated into their fractions in the crude distillation tower. The products are liquefied petroleum gas (essentially propane gas and butane), naptha (which is then processed into gasoline), kerosene/aviation turbine fuel, diesel oil and residual fuel oil. These products are further processed in thermal and catalytic cracking units (alkylation and reforming processes) to produce higher quality and stable products. The degree of refining complexity will dictate how many of the secondary processes are present. The simplest refineries (topping refineries) use distillation to separate gasoline or lube oil fractions from crude, leaving further refining of their residuum to other refineries or for use in asphalt. More sophisticated refineries will have thermal and/or catalytic cracking capabilities, allowing them to extract a greater fraction of gasoline blending stocks from their crude. The largest refineries are often integrated with chemical plants, and utilise the full range of catalytic cracking, hydroprocessing, alkylation and thermal processes to optimise crude utilisation. These processes are discussed briefly below.

14.2.1   Desalting, Crude Distillation and Vacuum Distillation

The first step in refining is to rid the crude oil of salt and sediment. This is done in a desalter. Next step is to separate the hydrocarbon fractions according to their boiling points. This is done by preheating the desalted crude and then separating them first in the atmospheric distillation unit. This separates the light fractions such as propane and gasoline from the heavier fractions such as heavy gas oil and heavy crude residue. The vacuum distillation unit takes the heavy crude residue and then distils them again under vacuum to prevent the decomposition of heavier hydrocarbons. The gas oil is sent to the cracker. The heavy distillates are converted to lubricating oils and asphalt.

14.2.2   Thermal and Catalytic Cracking

Some of the fractions from the crude distillation unit undergo further processing in thermal and catalytic cracking units where larger hydrocarbon molecules from the middle distillate and gas oil are broken into smaller molecules either thermally or catalytically. Catalytic cracking is one of the most widely used refining processes. It produces large amounts of sour water which contains BOD, ammonia, phenols, cyanides and sulphides. Hydrocracking (Isomax) is cracking in the presence of hydrogen. The end products are jet fuel and gasoline.

14.2.3 Hydrotreating

Sulphur, heavy metals, oxygen, halogens and nitrogen are impurities found in crude oil. These need to be removed because they poison or deactivate catalysts as well as cause emission problems. Hydrotreating converts unsaturated hydrocarbons (such as ethylene) into saturated hydrocarbons (such as ethane). Hydrogenation converts organic sulphur into hydrogen sulphide and is then recovered as elemental sulphur. Nitrogen is converted into ammonia and can be recovered as fertilizer feed.

14.2.4 Reforming

Crude distillation unit fractions have very low octane rating. The reforming process boosts the octane rating of hydrocarbons by rearranging the molecules. In the reformer, the same number of carbon atoms is kept but the process rearranges them into a different molecule. For example, cyclic hydrocarbon is reformed into aromatic hydrocarbons which have a higher octane rating. The reforming process uses catalysts such as Rhenium and Platinum and generates hydrogen gas.

14.2.5 Alkylation and Polymerisation

Smaller hydrocarbon molecules are combined into larger high octane gasoline feedstock. Catalysts are hydrofluoric and sulphuric acid.

14.2.6 Coking

Coking is a process that produces useful forms of coke from heavy residue products that are generated from other refining processes, such as vacuum distillation. The different coking processes include delayed coking, fluid coking and Flexicoking (a proprietary process developed by Exxon Research and Engineering).

14.2.7 Blending

This is the last phase where various feedstock are blended to achieve the desired end product characteristics.

Other processes are Lube oil production and asphalting.

A flow chart of a complex refinery is shown in Figure 14.3.

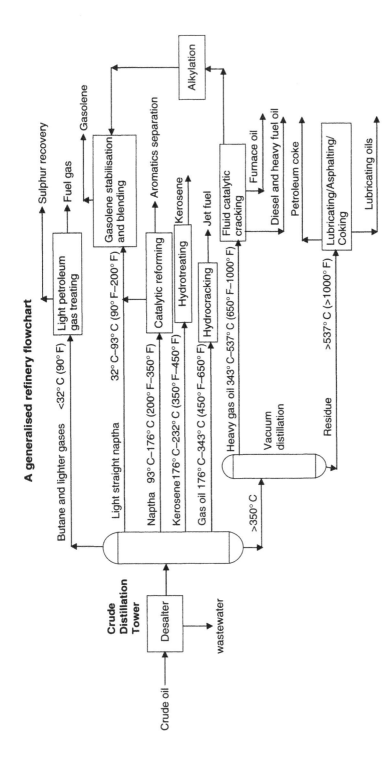

Figure 14.3   Flowchart of a complex refinery

14.3 Water Usage

The main purpose of water use in refineries is to transfer heat. Boiler feed-water makeup and cooling water account for 40–45% of water usage in refineries [4]. Other uses of water are as desalter wash water, quench water, drinking, sanitary and fire water. A hypothetical refinery's water balance is shown in Figure 14.4.

European Union benchmarks for total water usage is given as 0.1–4.5 m^3/ton of crude oil (3.6–162 US gal./barrel) and freshwater usage is given as 0.01–0.62 m^3/ton of crude oil (0.36–22.3 US gal./barrel)[5]. In the US oil refining industry the water usage ratios quoted by one source is 1.8–2.5 m^3/ton of crude oil (65–90 US gal./barrel)[4].

How much water is used in the oil refining industry?

From a water usage point of view, if roughly 2 m^3 of water is used to process one ton of oil and 0.3m^3 of wastewater per ton of crude is generated then this equates to 25.8 million m^3 (3.5 billion U.S.gallons) of water is used each day by the whole refining industry to process crude oil and in the process generates 3.5 million m^3 of wastewater.

14.3.1 Cooling Water Systems

Evaporative losses account for the bulk of water and energy losses in oil refineries. Cooling water systems play a very important role in rejecting heat from process streams. A cooling water system may consist of once-through (sea water), tempered closed loop cooling and open recirculating systems. In oil refineries air fin coolers are used upstream of water cooled heat exchangers since the process temperatures are higher than atmospheric dry bulb temperatures. Cooling water normally does not come into direct contact with process fluids and therefore is relatively clean. However heat exchanger leaks can dump oil to the cooling water system resulting in increased blowdown and generation of wastewater. Given the large cooling water requirements, it is a natural target for substitution of potable water. Chapters 5 and 6 provides more details on cooling water systems.

14.3.2 Steam Systems

The primary source of steam in the oil refining industry is utility boilers and heat recovery steam generators (waste heat boilers) and lately cogeneration plants. Around one-third of the steam is generated from waste heat boilers.

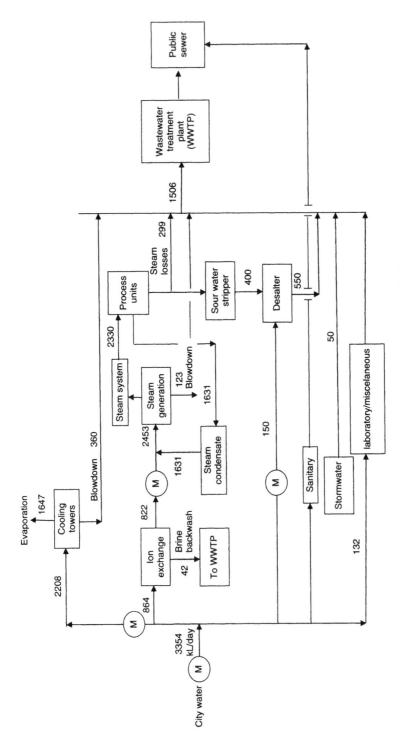

Figure 14.4   A water balance of a hypothetical refinery

The bulk of the steam is used in processes such as

- atmospheric distillation
- vacuum distillation
- catalytic hydrotreating
- reforming
- alkylation
- driving steam turbines.

In the process, large amounts of steam condensate is generated. Steam pressures are generally around 4020 kPa (600 psi) or less. In the US refining industry, 62% of industry's total capacity is less than 2027 kPa (300 psi) pressure [6].

Out of a refinery's total steam use about 40–60% is used directly in the refining process and therefore not available as returned condensate. Steam condensate recovery varies from 30% in older refineries to 70% in well maintained and in newer refineries. Water is lost through

- steam and condensate leakage
- dumping of oil-contaminated steam condensate
- non-recovery of condensate from tank farm areas due to low flows and long distances
- poor steam trap maintenance
- poor blowdown control
- venting to remove non-condensable gases from steam systems.

A total energy balance using energy pinch analysis is one way of identifying and rationalising energy and steam demands. Known as *Pinch Analysis* these are now used by oil refineries to optimise water and energy demands. Steam systems are covered in Chapter 7.

14.3.3   Wastewater Systems

As mentioned earlier wastewater generated per ton of crude ranges from 0.01 to 0.62 m^3/ton of crude (0.36–22.3 U.S. gal /barrel). Wastewater is generated from a number of process units. The process units that contribute the most wastewater are crude distillation, fluid catalytic cracking and catalytic reforming. A well designed and operated refinery could generate only 0.4 m^3/ton [7].

Refineries produce four types of wastewater. These are:

- Process water (desalter wash water, sour water)
- Surface water runoff (storm water)
- Cooling water, boiler water blowdown and pretreatment of ion exchange units
- Sanitary wastewater

A typical breakdown of wastewater sources is given in Table 14.1. Table 14.2 shows the contaminants in refinery wastewater.

The wastewater is treated in API gravity separation, dissolved air flotation units, biological treatment and clarification before discharged to the public sewers, to other receiving waters or preferably reused. Figure 14.5 shows a schematic of a typical wastewater treatment layout.

14.3.4   Energy Usage

The petroleum industry is a very large energy user. Energy usage is a function of type of crude processed, the complexity of the refinery as well as

Table 14.1   Refinery wastewater sources

Source	% of Total effluent
Cooling tower blowdown	20–35
Lost condensate	15–30
Desalter water	12–20
Excess sour water	10–20
Contaminated rain water	6–8
Boiler blowdown	4–8
Pretreatment	2–5

Table 14.2   Typical refinery wastewater quality and effluent limits

Water pollutant	Wastewater after pretreatment [4, 5]	Effluent limits maximum value [5]
pH	7–10	6–9
Biochemical oxygen demand (BOD)	150–400	30
Chemical oxygen demand (COD)	300–700	150
Total Suspended Solids (TSS)	2–80	30
Phenol	20–200	0.5
Oil	100–300*	
	5000**	10
Sulphide	5–15	1
Total Nitrogen	25–50	10
Benzene	1–100	0.05
Benzo (a) pyrene	<1–100	0.05
Total Chromium	0.1–100	0.5
Lead	0.2–10	0.1

All Units expressed as mg/L except pH.
* in desalter water
** in tank bottoms

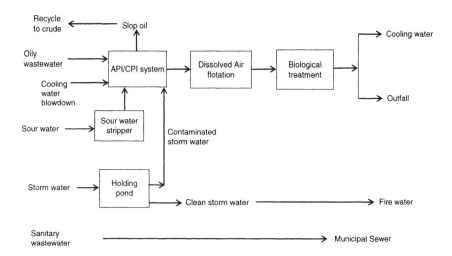

Figure 14.5   Refinery wastewater treatment flow schematic

other factors such as capacity utilisation, age of the refinery and operation and maintenance practices. The largest energy users in a refinery are the crude distillation unit, hydrotreater, reforming and vacuum distillation processes.

A refinery can purchase electricity or produce on site from cogeneration plants. Fuels used are usually by-products of the refining processes such as refinery gas and coke. Natural gas is purchased.

Given that utilities such as steam and cooling water systems play a major role in the refining processes, water conservation and reuse is closely linked to reducing utility costs.

Total energy consumption benchmark can vary from 470 to 1500 kWh per ton of crude oil [5].

14.4   Water Conservation Opportunities

Water conservation opportunities in oil refineries are not dissimilar to other manufacturing plants. In assessing the reuse options it would be helpful to be mindful of water-quality criteria and develop a matrix as shown in Table 14.3

Some options for water conservation are given below:

- substitute sewage treatment plant effluent as cooling water make-up
- reuse stripped sour water as desalter washwater
- investigate reuse of process water as cooling water make-up
- investigate recycle water as cooling tower make-up
- install cooling tower automatic blowdown control system
- improve recovery of steam condensate
- minimise boiler blowdown

Table 14.3 Water-quality matrix

Effluent Source	Type of pollutant and degree of contamination			
	High TDS	Low TDS	TSS	Oil
Cooling water				
Controlled blowdown		X	X	
Other uses and losses		X	X	X
Utilities effluents				
Softener/DI plant	X			
Boiler blowdown		X		
Condensate				
Stripped sour condensate		X		X
Other process condensate		X		X
Other non-process condensate		X		
Process water				
Desalter water	X			X
Other process water		X	X	X
Sanitary wastewater after treatment		X		
Ballast water				
Salt water	X		X	X
Storm water				
Process area		X	X	X
Other areas		X	X	X

- install boiler blowdown control systems
- use boiler blowdown as cooling water make-up
- retrofit amenities blocks
- establish leak detection and repair programme
- prevent solids and oily waste from entering the drainage system
- institute dry sweeping instead of washdown to reduce wastewater volumes.

Case Study: BP Refinery, Kwinana, Western Australia – Water Efficiency Initiatives

Openeed in 1955 BP Refinery Kwinana is Australia's largest oil refinery processing 19 100 metric tons (140 000 barrels/day) of crude oil. In 1997 it initiated a water reuse and minimisation programme with three main objectives:

- to minimise water use
- to maximise water reuse in refinery processes, either after or before treatment;
- to use low quality water (bore water) in place of potable water where practical.

The approach adopted involved four main steps:

Step 1 – Carry out a detailed analysis of the costs of wastewater treatment at the wastewater treatment plant.

Step 2 – Develop a detailed water balance.

Step 3 – Set targets for potable water usage, bore water usage, total water usage, flow to the wastewater treatment plant, water efficiency and percentage of condensate returned to the Refinery systems.

Step 4 – Examine all areas in the Refinery to determine where low quality water could be substituted for high quality water, and identifying areas where process usage could be reduced and returns increased.

An innovative aspect of the programme was approaching water management with a whole refinery perspective. All areas were targeted in order to save as much water as possible. All employees were encouraged to discuss and put forward ideas on water conservation, recycling or reuse. Monthly meetings were held within the refinery to discuss water minimisation. The initiatives are shown in Table 14.4.

14.4.1 Reduction in water consumption

The water reuse and minimisation programme results are given in Table 14.5.

Table 14.4 BP Refinery Kwinana water minimisation initiatives

Initiative	Water saving
Recyle process water in residue cracker unit	200–300 m^3/day.
Steam system – steam trap maintenance program	Improve condesate recovery through reducing leaks in steam condensate system.
Alky cooling tower make up	Replace with process water
In crease process water return	50% increase in process water return
Modifications to operating procedures	Reduce process water use.
Fire water system	Improved maintenace practices
Lube oil refinery	Increased condensate recovery
Reduce hydrocarbon spillage	Reduced process water usage
Stripped sour water	Reuse stripped sour water in desalters

Table 14.5 Water reuse and minisation programme results

Metric	Baseline year		% reduction
	1996	2003	
Total water use m³/day	7250	4686	35%
Scheme water use (potable water use) m³/day	6152	2006	67%
Wastewater flow, m³/day	5258	3386	36%
Condnesate return	32%	49.9%	17.9%

Adapted from Department of Environment and Heritage Canberra – *BP Refinery Kwinana – Cleaner production Initiative* and BP Refinery Water use and minimisation, Kwinana, Western Australia, 2003

References

[1] Energy Information Administration. *World Refining Capacity.* www.eia.doe.gov.

[2] Energy Information Administration. *Non-OPEC Fact Sheet.* http://www.eia.doe.gov/emeu/cabs/nonopec.pdf.

[3] Office of the Secretary of Energy, State of Okalahoma. Okalahoma Refinery Report Volume 1: *Challenges and Opportunities – A Study of the Okalahoma Refining Industry.* www.marginalwells.com/MWC/MWC/2005_Refinery_Rpt_Vol_1.pdf. 2005.

[4] U.S. Department of Energy. Industrial Technologies Program. *Water Use in Industries of the Future.* Washington D.C. July 2003.

[5] IPPC (Integrated Pollution Prevention and Control). Reference Document on Best Available Techniques for Mineral Oil and Gas Refineries. European Commission, Integrated Pollution and Control Bureau, Joint Research Center, Seville Spain. 2003.

[6] US Department of Energy, Office of Energy Efficiency and Renewable Energy. *Steam System Opportunity Assessment for the Pulp and Paper, Chemical Manufacturing, and Petroleum Refining Industries* – Main Report. Washington. October 2002.

[7] World Bank Group. Petroleum Refining. Pollution Prevention and Abatement Handbook. 1998.

Chapter 15

Laundries

15.1 Introduction

Commercial, institutional and industrial laundries are designed to handle a wide range of fabrics in large quantities. These facilities can use large quantities of water to clean hotel and healthcare linen, all manner of uniforms, mops, industrial mats and carpets.

Water is the universal solvent for the removal of soil and odour. However water quality can impact on cleaning efficacy. Calcium and magnesium hardiness affects the soaping ability of the water. Similarly the presence of iron and manganese can spot linen. Therefore water quality and water consumption is of importance to this industry. Water quality aspects are discussed in Chapter 2.

In commercial laundries labour accounts for close to 50% of a laundry's operating costs. After labour the next highest costs are for energy, water, chemicals, linen replacement and maintenance. Given the high proportion of labour related costs, gaining a marginal increase in productivity increases the launderer's bottom line significantly. Productivity improvements have been achieved through automation of the whole laundry process including collection, sorting, washing and drying, ironing and folding, packing and delivery of the finished goods whilst maintaining hygiene and cleanliness standards.

In a typical laundry, water related costs could be one third the utility costs. A commercial laundry can use as much as 400–500 m^3/day (105 680–132 100 US gallons/day) and over 85% of this water is used in the washing process. The other uses are for steam generation, cooling and amenities. Close to 70–90% of this water is then discharged to the public sewer. Commercial laundries not only discharge significant proportion of the incoming water, the effluent also may contain toxic and flammable chemicals that a public utility is incapable of removing before they are discharged to the environment. For this reason, environmental protection authorities and water utilities are also applying pressure on them to improve their effluent quality.

Given the important role of chemicals in the washing process, chemical suppliers have also responded by developing more environmentally friendly wash formulations such as low temperature and low pH wash formulations, the use of peroxide bleach formulations instead of chlorine bleach and minimizing the use of surfactants such as alkyl phenol ethoxylates (APE). These product changes have produced water and energy savings as well as environmental and occupational health and safety benefits.

15.2   Industry Structure

The laundry market consist of

- large commercial laundries and textile leasing companies
- hospital laundries
- large On-Premise Laundries (Large OPLs)
- small On-Premise Laundries (Small OPLs).

15.2.1   Large Commercial Laundries

The large commercial laundries and textile leasing companies are geared to provide the linen and cater to the heavy work wear from food-processing and automotive industries; hospital and food catering and restaurant industry linen and industrial wipers, carpets and mats. A characteristic is the presence of solvents, oil and grease and heavy metals in the soiled linen. Solvents such as creosote are occupational health hazards.

These laundries have capacities to process 3–50 tons of dry linen per day [1]. The main washing equipment consists of large washer extractors and continuous batch washers (CBWs). The laundries own the linen.

15.2.2   Large Hospital Laundries

Hospital laundries process bed linen (the vast majority), theatre clothing, patient clothing and theatre packing. Equipment and facilities are similar to large commercial laundries. One particular aspect of this linen is the presence of blood, pus, faeces, urine, pathogenic bacteria, food and medicine in the soiled garments.

Equipment capacities are around 3–30 tons of dry linen [1]. Laundry equipment consists of CBWs and washer extractors and laundries own the linen.

15.2.3   Large and Small On-premise Laundries

Large on-premise laundries (OPLs) are found in food-processing plants, hotels and hospitals. Daily processing capacities are 0.5–3 tons of dry linen [2]. Small OPLs are found in nursing homes, motels and retirement villages. Coin-operated laundries also can be included in the small OPL sector. These have loads of less than 500 kg/day. Equipment in both cases are washer-extractors.

15.3   Types of Laundry Equipment

The objectives of any washing process are

- to achieve maximum dirt and stain removal
- to minimise detrimental effects on the fabrics (mechanical and chemical damage)
- ensure hygiene standards are met (microbial contamination)
- the look and feel is enhanced (whiteness, softness and fresh and clean smell).

The washing process consist of the following stages:

1. Pre-wash (break wash) – removes loose soil and blood.
2. Wash – disinfection is carried out in this step. Alkaline pH of 11 and above is maintained.
3. Bleach – removes stains, enhances whiteness and provides chemical disinfection.
4. Rinse – removes suspended solids and any chemicals remaining.
5. Sour – reduces the pH of the water to neutral by the addition of acid to minimise further damage to textiles and reduce skin irritations.
6. Finishing – reduces moisture by extracting water from the textiles through centrifugal action.

A combination of factors are controlled to achieve the objectives in each washing process and these are

- Optimal Water usage
- Mechanical energy
- Heat
- Optimal chemical usage
- Time
- Number of washing operations.

A deficiency in one of the factors needs to be compensated by the other factors. Adjustments are made to the type of fabric and soiling. For example, in hospital laundries, blood can get fixated if the washing temperature is above 35°C.

The degree of mechanical action is managed by specifying the machine load for the particular machine. In washer extractors a typical ratio is 10 L for every kg of load. Overloading leads to reduced soil removal. Underloading leads to utilities being wasted.

Heat is required to make the oils and fats soluble in the water, for disinfection and for activation of bleaching chemicals. Higher the temperature the more efficient the soil removal process is; however, high temperatures also increases the cost of energy production. Regulatory or national standards such as the Australian and New Zealand Standard AS/NZS 4146:2000 : *Laundry Practice* specify a minimum temperature of 65° C for 10 minutes or 71° C for 3 minutes to kill pathogenic microrganisms [2]. In countries such as Germany, Austria and Belgium the thermal disinfection temperatures could be as high as 90° C and maintained for 10 minutes.

Chemicals are essential for washing and are added at every stage of the washing process and range from:

- alkalies (e.g. caustic, metasilicate) – increases pH to facilitate break-down in proteins, soil removal and increases the efficacy of bleaching chemicals.
- surfactants (like linear alkyl ethoxylates) – these range from nonionic, anionic and cationic surfactants. The nonionic and anionic surfactants increase the soil removal capabilities and cationic surfactants are applied as fabric softeners.
- builders such as phosphate and chelating agents inactivate hardness minerals and thereby increase the cleaning efficiency of the surfactants. Phosphates (are being phased out due to environmental impacts on the receiving waters).
- bleaching chemicals such as hypochlorite, peroxides, perborates and percarbonates. Hypochlorite usage is decreasing since the byproducts are not biodegradable and have carcinogenic properties.
- Acids such as phosphoric and citric are used to neutralise the alkaline pH.

Timing can vary from 2 minutes for a rinse to 10 minutes for the wash cycle. The number of washing operations performed depends on the type of soil and fabric to be washed. To achieve these objectives two types of laundry equipment are used in laundries. These are

- Washer extractors
- Continuous batch washers (Tunnel Washers).

They both accomplish the same goal. Successive baths are used to wash, rinse and finish the goods.

15.3.1 Washer-Extractor

A washer-extractor does all of the washing in one cylinder similar to the domestic variant. The load stays in place, and baths are changed by draining and refilling the single cylinder. An extraction cycle spins water out of the goods and prepares them for drying. Washer-extractors range in sizes from 16 to 363 kg (35 to 800 lbs). They can be either top loading or more commonly of the front loading or side loading types. Front loading machines use much less water and energy than the top loading models. Washer extractors conserve water in two ways. Intermediate extracts reduces the number of rinse water cycles. Some designs reduce water usage by injecting the water from below and tumbling the linen. These are reportedly more efficient that spray rinses. Other water conservation opportunities in these machines include proper scheduling, collection of rinse water in external tanks and other measures which are described below. Around 30% of water can be saved by reusing the final rinse water [3]. Modern washing extractors have programmable logic control and card systems to optimise water and chemical usage. Extraction speeds can be varied according to the linen washed thus saving on energy costs.

15.3.2 Continuous Batch Washers or Tunnel Washers

CBW systems are the work horses of the large industrial laundries or more commonly called *tunnel washers*. They were first invented by the German engineer Erich Sulzmann. The linen is moved by an internal 'auger' through a number of internal compartments. Each compartment requires a certain washing temperature. This is achieved by direct injection of steam to heat the water within the machine. Modern CBWs come with internal recycling options. Fresh water enters from the opposite direction i.e. the chamber with the cleanest linen. The water then cascades to the other chambers and then discharged. The use of counter current washing reduces the water and steam usage by 60–70% compared to the washer extractors. It also saves on detergents and chemicals used. At the end of the tunnel the linen is removed automatically and sent to dryers.

Figure 15.1 A photo of a typical CBW System

Courtesy of Sdyney Water.

Case Study: Parramatta Linen Service installs 3 new CBWs and halves its water usage.

Parramatta Linen Service (PLS) a NSW government owned laundry is the largest laundry in Australia. It processes 250 tons of linen a week and supplies 70% of the NSW Health's requirements for hospital linen, ward and sterile theatre linen and surgical garments.

PLS was consuming 12 L of water/kg of dry linen. It recently replaced the 3 CBWs with new CBWs and reduced its water usage to 6.4 L/kg of dry linen. Thus saving $250\,m^3$ of water per day and $78,000 per year.

Source: Media release, NSW Government, 26 February 07 and personal communications.

15.4   Benchmarking

In the US, a study conducted by the US Laundry environmental stewardship program of 500 laundries established that the overall water usage is 19 litres per kg of dry linen (2.28 US gal/lb) whilst dedicated linen plants use around 9.2–10 L/kg (1.1 - 1.2 US gal /lb) [4]. Average energy usage for the 500 respondents is estimated at 7.4 MJ/kg (3200 Btu/lb).

European values for tunnel washers are in the range of 8–12 L/kg (0.96–1.44 US gallons/lb) [1]. For washer extractors, Sydney Water has estimated the best practice water usage values and these are shown in Table 15.1 [3].

Besides heating of wash water, steam is used for heating of air in tumble dryers and ironers. For tunnel washers steam consumption is in the range of 0.4–0.5 kg/kg [1]. For washer extractors, steam consumption is in the order of 0.6–0.8 kg/kg of linen (0.27–0.36 lb of steam/lb of dry linen) [3].

In the UK, according to a survey carried out by the Energy Efficiency Office in 1993 established that the mean energy consumption was 2.66 kWh/kg [5]. The survey also established a good practice target of 1.96 kWh/kg.

Table 15.1 **Best practice water usage for washer-extractors**

Rating	Water usage without reuse		Water usage with reuse	
	Litres/kg Linen	US gal./lb linen	Litres/kg Linen	US gal./lb linen
Good	17–22	2.0–2.6	12–15	1.4–1.8
Fair	22–26	2.6–3.1	15–18	1.8−4.8
Poor	>26	>3.1	>18	>4.8

Courtesy of Sydney Water. Water Conservation Best Practice Guidelines for Hotels. Sydney. December 2001.

15.5   Water Conservation

Water conservation opportunities for laundries can be classified as follows:

- Get senior manager commitment. Refer to Chapter 3 on how to plan for water conservation.
- Install sub-meters and track daily water- and energy-usage to identify abnormal behaviour in machines or when processing a particular type of linen.
- Improve maintenance practices such as fixing leaks in water systems. Generally a 3% reduction in water usage is realizable [6]
- Improve scheduling of washing equipment and modify wash programs to suit the linen.
- Do not under load washers. A 10% under load results in 10% wastage of water and energy [6].
- Reduce rewash rates to a minimum.
- Change chemical ingredients to low pH and low temperature formulations.
- Purchase water efficient washer – extractors and CBWs with internal recycling.
- Install water efficient fixtures in amenities blocks.
- Replace defective steam traps and recover steam condensate.
- Repair live steam and condensate leaks.
- Ensure that the heat reclaimers are operating properly. Install temperature gauges to ensure that the unit is working properly to recover heat.
- Improve dryness of steam to processing areas.
- Recover vented flash steam. Refer to Chapter 7.
- Lag steam and condensate lines. Poorly insulated lines are a safety hazard and gives out heat through radiation.
- Ensure that the boiler blowdown practices are in accordance with national standards. High dissolved solids lead to scaling of boiler tubes.
- Practice on-site water reuse.
- Do not wash above the recommended temperature settings.

Water-reuse options are discussed in greater detail below.

15.5.1   Water and Energy Reuse

Laundry washing consists of discharging a very high proportion of the water it uses to the sewer. Moreover the water is hot and frequently needs to be cooled to below 38° C before discharging to the sewer. By recovering the waste heat using a heat exchanger as much as 60% of the heat can be recovered. The wastewater quality is frequently at a high pH. If the alkalinity can be recovered then the chemical costs can be reduced. Table 15.2 shows the wastewater quality from an industrial and two hospital laundries.

Table 15.2 Wastewater quality from industrial and hospital laundries

Parameters	Industrial laundry	Hospital laundry	Hospital laundry 2	License discharge limits to sewer
pH	8.26	11.4	11.6	7–10
Conductivity, μS/cm	640	808	1000–2000	NA
Total dissolved solids, mg/L	420	456	800	10,000
Total suspended solids, mg/L	69	71	66	600
Total hardness as mg/L CaCO₃	44	68	53	NA
Total alkalinity, mg/L CaCO₃	128	302	375	NA
Total oil and grease, mg/L	24	26	25	110
Phosphate, mg/L	3.43	10.8	167	50
BOD, mg/L	262	50	44	230

A consideration for installing water-reuse plants are that they are space limited and therefore the equipment needs to have a small footprint.

As Table 15.2 shows the major contaminants are high pH, suspended solids, oil and grease and BOD. Industrial laundries would have a higher concentration of oil and grease since they wash industrial carpets and mats, work clothes from a variety of industries such as from the automotive sector. Petroleum hydrocarbons could be an issue in some cases. Besides this laundry wastewater contains lint, hair and heavy metals.

The common wastewater external reuse strategies are

- Partial reuse of water
- Total reuse of water
- Ozone technology.

15.5.2   Reuse for Wash Water Quality

Partial reuse of water systems are designed to reuse of water as wash water rather than as rinse water. These systems typically do not reduce total dissolved solids and as such cannot be used in the rinse cycles.

Retrofit conventional washers with a holding tank to capture final rinse water and reusing in the wash cycle has been a common practice. It is estimated that this option has 30% water saving [4]. Since only the rinse water is captured, these systems typically have high initial costs and long payback periods [8].

Another option is to treat the total effluent and use it only in the wash water cycle. The effluent from the collection pit is first sent to an equalization tank and from there to a static screen or lint shaker. The lint shaker screen removes particulates. From here the water is pumped to a clarifier or dissolved air flotation unit (DAF) with coagulants and polymers injected online. The chemicals breakdown oil and grease and trap remaining suspended solids. However no dissolved solids are removed. From the tank the water is sent to a reclaimed water storage tank. The water is used for hot water applications only. The rinse and bleach cycles use fresh water. The total dissolved solids levels are monitored and controlled within allowable limits by diluting with freshwater. Another system marketed by Wastewater Resources Inc known as the Aquatex 360 uses backwashable pressure filters to remove 90–95% of the suspended solids [9]. Manufacturers claim to save 50–65% of the water and energy [8]. Chemical savings are achieved because the reclaimed water is at a pH of 9.5–10.5 compared to towns water pH of 6.5–7.5. Thus a reduced amount of caustic is required to boost the pH to 11.5.

Another manufacturer of such systems is Ecolab Inc. The case study below gives a description of their system.

Case Study: Illawarra Linen Service, Wollongong, Australia

Illawarra Linen Service is a NSW government-owned hospital linen service in Australia. It has a daily washing load of 15 tons per day and used to consume over $260\,m^3$/day of water.

It installed and Aquamiser unit from Ecolab Australia that allowed it to reduce its water consumption by $47\,m^3$/day. The Aquamiser removes sand, dirt and lint down to $25\,\mu m$. The reclaimed water is used in the washing cycles. The Energy Optimiser allows it to recover heat reducing the amount of natural gas required to fire the thermal oil heaters.

Adapted from: Sydney Water. The Conserver. Issue 9. December 2005.

15.5.3 *Reuse for Rinse Water Quality*

In total reuse of water suspended solids as well as dissolved solids are reduced. To reduce total dissolved solids most systems use nanofiltration or reverse osmosis membranes. However, due to the high suspended solids and oil and grease loadings nanofiltration and reverse osmosis systems cannot be used without pre-treatment that frequently includes microfiltration or ultrafiltration. The need for extensive pre-treatment makes the use of membrane systems cost prohibitive. Chapter 8 gives an explanation of membrane systems.

The effluent known as permeate can then be used as wash water or as rinse water. Generally 80–85% of the water can be recovered. The reject water is sent to the sewer.

The Vibratory Shear Enhanced Process (VSEP) from New Logic Research Inc. is a membrane system that does not require extensive pre-treatment apart from a shaker screen to remove lint. It is capable of operating to produce

Table 15.3 Permeate and reject water analysis from a VSEP nanofiltration membrane

Parameter	Effluent water	Filtrate ex Nanofiltration membrane	Reject stream to sewer
pH	11.6	11.0	11.3
Conductivity, μS/cm	1000–2000	320	3,000–4,000
Total dissolved solids, mg/L	800	134	1000–2000
Total suspended solids, mg/L	66	Nil	300
Total hardness as mg/L CaCO₃	68	<1	
Phosphate mg/L	167	<1	470
Chloride, mg/L	61	14	75
Iron mg/L	0.32	<0.1	
Total alkalinity as CaCO₃ mg/L	375	68	760
Total grease mg/L	25	<1	57
BOD	44		

Courtesy of the Water Management Group Pty Ltd.

washwater quality water as well as rinse water quality water. Table 15.3 shows the results achieved from a VSEP nanofiltration membrane trial. Refer to Chapter 8 for a discussion of this system.

The nanofiltration membrane is not able to completely remove monovalent ions and for this reason the alkalinity is only partially reduced. Figure 15.2 shows a schematic of the wastewater system.

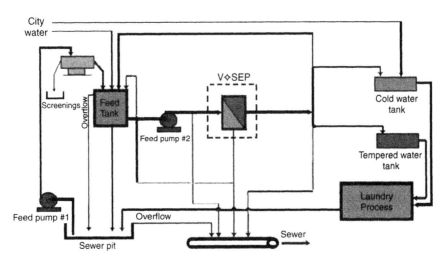

Figure 15.2   A schematic of a water-reuse system

Courtesy of New Logic Research Inc.

Case Study: Hospital Central Services Association, Seattle, USA

The Central Services Association processes the laundry of 11 hospitals amounting to 9 090 000 kg (20 000 000 lbs) of goods per year. The laundry operates for 14 hrs of the day and operates for 364 days of the year. It uses 395 m^3/day (104 US gal./day). After careful analysis the laundry installed a VSEP ultrafiltration system to recover 80% of the effluent water. At 80% recovery the operational savings are

The calculations are shown below:

Water and wastewater charges before reuse	US$218 000

Costs

VSEP power consumption (30 kW @ US$0.38 kW)	US$ 6 600
System maintenance and cleaning	US$ 9 838
Water and wastewater charges after reuse	US$ 37 400

Additional Savings

Heating savings (from 60° F to 100° F), 256 Therms/day @US$0.32/therm	US$ 35 438
Detergent savings	US$ 20 000

Total savings/yr = US$218 000 − 6 600 − 9 838 − 37 400 + 35 438 + 20 000 = US$219 600/yr

Payback = 11 months

Adapted from: Paschke P., Hill S., Lawson S. *At the Vanguard of Commercial Water Conservation. Seattle Public Utilities. Seattle Washington* and New Logic Research Inc. *Industrial Laundry Wastewater Treatment – a cost-effective and environmentally sound treatment.* January 1999.

15.5.4 Ozone Disinfection

The disinfection properties of ozone are well known. Ozone is now being applied in lightly soiled applications as a disinfectant and bleaching product instead of hypochlorite.

Ozone (O_3) is a colourless to blue gas with a pungent odour and a powerful disinfectant. It is denser than oxygen with a density of 2.14 kg/m^3 compared to 1.43 kg/m^3 at standard temperature and pressure. O_3 has an oxidation potential far greater than the hypochlorous acid (HOCl) of − 2.07V as against −1.49V. As a disinfectant, O_3 is 150% more potent and 3000 times more reactive than chlorine. It only requires as little as 1.5 mg/L of O_3 to kill bacteria and viruses. Similar to chlorine it also oxidises other soluble organic and inorganic contaminants such as tannins, iron and manganese.

However, O_3 is highly unstable and needs to be generated on-site. It has very low solubility and at temperatures above 43° C it is not soluble in water. Table 15.4 shows basic data on ozone. At pH levels below 7 the reactivity

Table 15.4 O_3 conversions and data

Parameters	
Density of O_3	2.14 kg/m³ at standard temperature and pressure
O_3 concentration in water	1 mg/L = 1 ppm O_3 = 1 g O_3/m³
O_3 concentration in air by volume	1 g O_3/m³ of air = 467 ppm
Ozone solubility in water at 1% concentration (12.8 g O_3/m³)	5° C–7.39 mg/L
	15° C–5.60 mg/L
	25° C–3.53 mg/L
Disinfection	bacteria and viruses – 1.5 mg/L
Oxidation – dissolved organic carbon	0.5–3.0 mg/L = 0.5–1.5 g/g COD
Pre-oxidation iron (Fe)	0.43 mg/L × Fe mg/L
Pre-oxidation manganese	0.88 mg/L × Mn mg/L

of O_3 is slow. However, at alkaline pH > 8, O_3 rapidly decomposes to the hydroxyl-free radicals which react very quickly with contaminants. Optimum pH for laundry applications are in the range 8–10.

O_3 is also a highly toxic gas and even at very low concentrations can be harmful to the upper respiratory tract and the lungs. Concentrations as low as 0.08 parts per million and exposed for 7 hours can bring about discomfort and an exposure of 50 ppm for half an hour can be lethal [10]. O_3 will also oxidise and corrode metals such as copper and its alloys. Only certain plastic materials and stainless steel are compatible with ozone applications. Best to consult a reputable ozone generator manufacturer. In poorly ventilated environments it is recommended to install an ozone monitor to monitor O_3 concentrations in the air.

In laundry applications, O_3 reduces water, energy and chemical costs. Water savings are achieved by reducing the number of rinses and some of the detergent chemicals. Although suppliers claim that water and sewer savings are as high as 50% general agreement is that this is more in the range of 15–20% [8]. Since O_3 works best at temperatures below 25°C (where a minimum dose of 3.0 mg/L is achievable) there are energy savings to be realised in the order of 40–60% as claimed by some manufacturers. There are claims that O_3 also reduces chemical costs by as much as 50%. However this depends on whether the washing formulations need to be changed to accommodate the low temperatures. Other benefits of O_3 are that the clothes are softer and clean smelling.

The efficacy of O_3 washing is dependent on how it is injected into the washer. The main methods of O_3 injection are

- bubbling O_3 gas directly into the washer
- direct injection
- supplying ozonated water through a supply tank.

Ozone dissolution in water follows the rules for solubility of gases and all of these methods depends on the four main factors that affect the solubility of O_3 in water are

- contact time between the gas and the water
- water pressure
- method of mixing of ozone in the water
- water temperature.

In the bubbling method, the efficacy of O_3 is dependent on how quickly the gas can dissolve during the cleaning cycle. Given the low solubility of O_3 and the short contact time of the cleaning cycle, the majority of the O_3 would be expected to flash off into the vapour phase. Moreover, it is difficult to precisely control the dosage of O_3 in the wash mix.

In the direct injection method, an eductor is used to suck the O_3 gas to the water supply line to the washer. This method too relies on having sufficient contact time to dissolve the O_3 gas. One way increasing the ozone and water contact time is to install a static mixer. Having sufficient contact time is crucial to achieve the desired O_3 concentration in the washer.

Dissolving the O_3 in a tank allows for sufficient time to achieve the desired concentration. The ozonated water can then be used in the washing process. This is preferred method if the site is not limited by space.

It is desirable to measure the ozone concentrations through a dedicated ozone monitor or an oxidation reduction potential monitor (ORP). Typical values to achieve 1.5 mg/L of O_3 ORP are in the order of 700–975 mV.

The ASTM Standard E 2406–04 [11] can be used to evaluate the laundry sanitisers and disinfectants for use in high-efficiency washing operations. Organisations such as the Australian Wool Testing Authority are able to test the claims of suppliers regarding the stain-removal capabilities of O_3.

References

[1] Association Internationale de la Savionnerie, de la Detergence et des Produits d' Entretien. *Environmental Dossier on Professional Laundry.* www.aise-net.org/downloads/AISE_pro_laundry_dossier.pdf. October 2000.

[2] Standards Australia. Australian/New Zealand Standard – Laundry Practice. AS/NZS 4146:2000. Sydney. 2000

[3] Sydney Water. *Water Conservation Best Practice Guidelines for Hotels.* Sydney. December 2001.

[4] Koepper K. LaundryESP data base bears more fruit. Industrial Launderer. www.ilmagonline.com. May 2006 issue.

[5] Energy Efficiency Best Practice Programme. *ECG 049 – Good Practice Guide – Energy Efficiency in the Laundry Industry.* Harwell, Oxfordshire, UK. 1995.

aerobic conditions over a 5-day period at a temperature of 20°C. The 5-day test equates to 2/3 of the total BOD.

Boiler horsepower – A unit of rate of water evaporation equal to the evaporation per hour of 34.5 pounds of water at a temperature of 2120°F into steam at 2120°F. One boiler horsepower equals 33 475 Btu/hr.

British Thermal Unit (Btu) – The amount of heat required to raise the temperature of one pound of water by one degree Farenheit; equal to 252 calories or 4.18 J. It is roughly equal to the heat of one kitchen match.

Bulking activated sludge – The term used to describe sludge in the secondary clarifier with poor settling characteristics. This is often due to a high proportion of filamentous bacteria and a lack of health floc–forming bacteria.

Carbohydrates – A class of compounds containing carbon, hydrogen and oxygen. General formula is $(CH_2O)_n$. Examples are sugars and starch.

COD (chemical oxygen demand) measures the total *organic* content that can be oxidised by potassium dichromate in a sulphuric acid solution and is expressed as mg/L.

Cogeneration – The simultaneous production of electrical or mechanical work and thermal energy from a process, thus reducing the amount of heat or energy lost for the process. Also known as combined heat and power (CHP).

Coliform bacteria are non-pathogenic microbes found in faecal matter that indicate the presence of water pollution; are thereby a guide to the suitability of water for potable use.

Colony-forming unit (CFU) is a measure of viable bacterial numbers. Unlike in direct microscopic counts where all cells, dead and living, are counted, CFU measures viable cells. A sample is spread or poured on a surface of an agar plate, left to incubate and the number of colonies formed are counted. The CFU is not an exact measure of numbers of viable cells, as a colony-forming unit may contain any number of cells.

Colour – A measure of dissolved substances in water that give it a coloured appearance. The measurement is in Hazen units. It is a comparison to platinum and cobalt salts. The drinking water standard is 15 Hazen units.

Condensate – Condensed steam.

Condensate pump – A pump that pressurizes condensate allowing it to flow back to a collection tank, deaerator or boiler.

Condenser – A heat exchanger in which a refrigerant or steam cools and then condenses from vapour to liquid.

Conductivity measured in microS/cm ($\mu S/cm$) is used as a proxy for TDS.

Continuous blowdown – The process of removing water, on a continuous basis, from a boiler to remove high concentrations of dissolved solids, chlorides and other products. Water is replaced by treated make-up water add to the condensate.

Dalton – The unit of measurement of molecular weight, named after the physicist and chemist John Dalton, founder of the atomic theory. The Dalton is used to denominate the molecular weight of very complex molecules and is defined as one-twelfth of the weight of the atom of the C12 isotope of carbon.

Deaerator – A device that uses steam to strip feedwater of oxygen, carbon dioxide and ammonia.

Dissolved oxygen – A measure of the oxygen dissolved in water. In aerobic wastewater treatment plants the oxygen levels can vary between 0.5 and 3.0 mg/L depending on the process. In cooling water and boiler water systems dissolved oxygen contributes to corrosion.

Drift Loss – This is the additional carry over of water that occurs from cooling when droplets of water escape from the cooling tower.

E-coli (Escherichia coli) is also one of the non-pathogenic coliform organisms used to indicate presence of pathogenic bacteria in water.

Effluent is the liquid, solid or gaseous product discharged or emerging form a process.

Evaporation rate – The rate at which water evaporates as part of the evaporative cooling process.

External Water Footprint – External Water Footprint is defined as the annual volume of water resources used in other countries to produce goods and services consumed by the inhabitants of the country concerned.

Floating matter is that which passes through a $2000\,\mu$ sieve and separates by floatation in one hour.

Food to mass ratio (F:M) – The ratio of the bacterial food, that is BOD load, and the mass of bacteria available to treat the BOD. An important control parameter for the activated sludge process. Also referred to as the sludge loading rate (SLR).

Flux – Flux is the quantity of material passing through a unit area of membrane per unit time. SI units are $m^3/m^2/s$ or simply m/s. Other non-SI units are $L/m^2/h$ (LMH) and m^3 per day. The US units are gal./ft^2/day (GFD).

Food Cover – Any transaction or sale, whether a cup of coffee or a multiple course meal.

Greenhouse gas emissions – Those gases, such as water vapour, carbon dioxide, ozone, which are believed to contribute to climate change.

Heat exchanger – A device used to transfer heat from one medium by either direct or indirect contact.

Heat Recovery Steam Generator – A type of wasteheat boiler that captures the thermal energy in cogeneration systems and produces steam.

Internal Rate of Return (IRR) – Internal Rate of Return is the discount or interest rate at which the net present value of an investment is equal to zero.

Internal water footprint – Internal water footprint is defined as the use of domestic water resources to produce goods and services consumed by the inhabitants of the country.

Kilowatt hour (kWh) – A unit of measure of electric supply or consumption of 1000 W over the period of one hour; equivalent to 3412 Btu.

Kilowatt refrigeration (kWr) – Kilowatt refrigeration as distinct from kW electrical.

Latent heat – The change in heat content that occurs with a change in phase without a corresponding change in temperature. Changes in heat content that affect a change in temperature are called sensible heat changes. The Latent heat of water is 2431 kJ/kg (1000 Btu/lb).

Life cycle cost (LCC) – Cost estimate of a piece of equipment over its entire life, including development costs, production costs, warranty costs, repair costs and disposal costs.

Make up – Water supplied to a cooling tower to replenish water lost through evaporation, bleed, drift and wastage.

Membrane – A thin barrier that allows some compounds or liquids to pass through, and impedes others. It is a semi-permeable skin of which pass through is determined by the size or special nature of the particles. Membranes are commonly used to separate particles and dissolved ions.

Mixed Liquor Suspended Solids (MLSS) – A measure of the solids content of an activated sludge reactor. Generally measured in mg/L.

Molecular Weight Cut-off (MWCO) – Molecular weight cut-off is the nominal molecule size rejected by a membrane. The units are expressed in Daltons.

MPN Index (most probable number) is used to report results of the coliform test for bacteria. It represents the number of coliform bacteria in the water.

Net Present Value – Net Present Value (NPV) is equal to the present value of a future returns, discounted at a marginal cost of capital, minus the present value of the cost of the investment.

Nitrification – The biological oxidation of ammoniacal nitrogen to nitrate. It is a two-stage process requiring two types of bacteria. *Nitrosomonas* sp. convert ammoniacal nitrogen to nitrite, then *Nitrobacter* sp. convert nitrite to nitrate.

Organic nitrogen is the nitrogen combined in organic molecules such as proteins, amines and amino acids.

ORP (oxidation–reduction potential) indicates the degree of completion of the chemical reaction by detecting the ratio of ions in reduced form to those in the oxidised form as variations in electrical potential form an ORP electrode assembly.

Overflow – The water which may accidentally flow over the basin of a cooling tower, or through the tower overflow pipe, due to a deficiency in the condenser water system.

Pathogenic bacteria are those micro-organisms capable of producing diseases in man, animals and plants such as *Legionella, Salmonella, Giardia, Listeria* and *Camphylobacter.*

Payback – The amount of time required for positive cash flows to equal the total investment costs. This is often used to describe how long it will take for the savings resulting from using water and/or energy efficient equipment to equal the premium paid to purchase the water (energy)-efficient equipment.

Payback period – Measures the length of time taken for the return on an investment exactly to equal the amount originally invested.

ppb – Parts per billion. Commonly considered equal to micrograms per litre (μg/L).

ppm – Parts per million. Commonly considered equal to milligrams per litre (mg/L).

ppt – Parts per trillion. Commonly considered equal to nanograms per litre (ng/L).

Refrigerant – The working fluid of the refrigeration system such as ammonia or other that absorbs heat from the evaporator (chiller) and then rejects it in the condenser.

Risk management – The process for identifying, assessing, monitoring and controlling the key factors that could affect the desired outcome of a maintenance procedure.

Sensible heat change – A change in heat content with a corresponding change in temperature.

Settleable solids are those in suspension that will pass through a 2000 μ sieve and settle in 1 hour under the influence of gravity.

Sludge age – Often referred to as the mean cell residence time. A measure of time the biomass is retained within the system. It is closely related to the F:M ratio and an important plant control parameter.

Steam ejector – A device that uses a relatively high pressure motive steam flow through a nozzle to create a low-pressure or suction effect.

Steam trap – An automatic control device that allows for the removal of condensate, air and carbon dioxide and other non-condensable gases, whilst keeping living steam in the system.

Suspended solids are those solids that can be removed by filtration.

Thermal conductivity – This is a positive constant k, that is a property of a substance and is used in the calculation of heat transfer rates for materials. It is the amount of heat that flows through a specified area and thickness of a material over a specified period of time when there is a temperature difference of one degree between the surfaces of the material.

Throttling – Regulating flow rate by closing a valve in the system.

Total dissolved solids (TDS) – A measure of the salinity of water, expressed in mg/L.

Total organic carbon (TOC) – A measure of the organic carbon content in an effluent. A useful indicator of the organic strength, particularly suited to on-line monitoring, expressed in mg/L.

Total solids equal the sum of suspended solids and dissolved solids.

Turbidity is the amount of suspended matter in wastewater obtained by measuring its light-scattering intensity and reported as NTU.

Turbine – A device that converts the enthalpy of steam into mechanical work. There are two types of turbines – condensing and backpressure turbines. Condensing turbines exhausts steam to sub-atmospheric conditions, where the steam is condensed. These are normally found in power plants. Backpressure turbines exhausts steam above atmospheric conditions. The exhaust steam is usually sent to a lower pressure system.

Variable Frequency Drive – A type of variable speed motor drive in which the motor is supplied with electrical power at frequencies other than standard 60 Hz through a converter.

Virtual Water – Virtual water is defined as the volume of water needed to produce a commodity or service.

Virus – Any large group of sub-microscopic infective agents capable of growth and multiplication only in living cells of a host. Examples are given in Chapter 2. Phages are used as a proxy for virus contamination.

Waste heat – Heat that is discharged from a mechanical process, wastewater or ventilation exhaust system that could be reclaimed for useful purposes.

Water footprint – The water foot print of a country is defined as the volume of water needed for the production of goods and services consumed by the inhabitants of the country. National water footprint consists of internal water footprint and external water footprint.

Wet bulb temperature – The lowest temperature that can be obtained by evaporating water into the air at constant pressure. The average yearly wet bulb temperature for Sydney is 14.9° C.

Appendix

Worksheets

Worksheet 1

Location of Water Meters

Account number	Meter number	Size/type of meter	Meter location

Worksheet 2

Month	Year	Total for all meters (kL/month)	Water usage charges ($)	Fixed water charges	Sewer usage charges ($)	Fixed sewer charges ($)	Trade waste charges ($)	Total charges	Monthly production units	Benchmark L/unit of product
January										
February										
March										
April										
May										
June										
July										
August										
September										
November										
December										
Total										

Worksheet 3

Water-Use Inventory

Item	Location	Number of units (A)	Flow rate		Operating time (D)	Total water used (kL/day) (1000 US gal./day)		Water saving [1–2] (kL/day) (1000 US gal./day)
			Current (B)	Water-efficient models (C)		Current (A × B × D) [1]	Proposed (A × C × D) [2]	
Taps	Guest rooms, amenities, kitchen, laundry	400	12 L/min (3.17 US gal./day)	6 L/min (1.6 US gal./day*)	15 mins/day	72 (19)	36 (9.5)	36 (9.5)
Showers	Guest rooms	400	15 L/min	9 L/min	12 mins/day (2 showers/ day)	81 (21.4)	48.6 (12.8)	32.4 (8.6)
Toilets	Guest rooms	300	11 L/flush	6/3 L/flush**	5 flushes/day	18.6 (4.9)	7.6 (2.0)	11 (2.9)
Wok stoves	Kitchen	2	8 kL/day	3 kL/day		16	6	10
Pre-rinse spray valves	Kitchen	3	19 L/min	6 L/min	5 hours/day	17	5.5	11.6

* US federal standard is 2.5 gal./minute
** US federal standard is 1.6 gal./flush

Check Sheet

Sydney
WAT≡R

Water saving check sheet for office buildings

Saving water in your business can reduce the stress on water resources, help protect the environment and save you money. This check sheet is intended to give facility managers of office building guidance on implementing actions to reduce water usage

Area	Question	Answer	Action
Amenities	Is the flowrate from the bathroom faucet less than 6L/min?	Yes ☐ No ☐	If no, install flow control to reduce water flow by over 50% (flow control will reduce flow to <6L/min in bathroom basins and <9L/min in kitchen sinks)
	Do you have dual flush toilets?	Yes ☐ No ☐	If no, where cost effective replace the cistern and pan with a 6/3L dual flush model
	Is the toilet flush volume greater than 6L?	Yes ☐ No ☐	If yes, reduce flush volume by adjusting cistern float arm, installing a displacement device, or adjust flush valve to shorten the period of flush. For 11L cisterns, reducing flush volume to 8L will still provide an adequate flush
	Are cistern rubber seals regularly replaced?	Yes ☐ No ☐	If no, cistern rubbers should be replaced approximately every two years before leakage occurs
	If you have a flusherette system, have you checked the flush timing?	Yes ☐ No ☐	If no, check that the toilets are not being over flushed. A timing of 3 seconds for the flusherette valves should be adequate
	Do you have cyclic flushing urinals?	Yes ☐ No ☐	If yes, adjust timing cycles and if possible reduce flush volumes, ensuring flushing occurs only during work hours. Replace with either manual or automatic sensor flushing which can reduce water use by 70%. Payback is typically less than 6 months
	Do you have automatic sensor urinals?	Yes ☐ No ☐	If yes, check that the sensor beam is correctly set so that the urinal only flushes when in use, however not before 2 mins of detection and no more than 10min between flushes. In very busy areas timing between flushes should be reduced to no more than 6 mins. During out-of-hours periods either disable the flushing timer or minimise flush to once every 4 hrs
	Do you have water-efficient showerheads?	Yes ☐ No ☐	If no, consider installing an AAA-rated showerhead that can reduce flow from to 9L/min or less, otherwise install restrictors in the tap assembly or existing showerheads
Cooling	Does the air-conditioning system run continuously?	Yes ☐ No ☐	If yes, where possible turn off air-conditioning units when cooling is not required
	Is your water cooling tower cleaned regularly (every 6 months or less)?	Yes ☐ No ☐	If yes, adopt a performance-based approach with your water treatment contractor to minimise the amount of cleaning required
	Is the floor area around the cooling tower constantly wet?	Yes ☐ No ☐	If yes, install anti-splash louvres
	Does water overflow from the water cooling tower to the drain pipe?	Yes ☐ No ☐	If yes, further investigation is required. There are several possible causes, such as incorrectly set ball float valves. Refer to Sydney Water's best practice guidelines for assistance
	Does the water cooling tower have a side stream filter that uses water for back flushing?	Yes ☐ No ☐	If yes, capture the bleed-off from the cooling tower in a holding tank and use this water for backwashing. Alternatively, consider using a cartridge filter which will further reduce water usage
	Is the conductivity meter cleaned every month?	Yes ☐ No ☐	If no, ask the water treatment contractor to clean the probe every month and re-calibrate it at least every six months. A poorly maintained probe will increase bleed-off and waste water
	Does your water cooling tower operate at optimum cycles of concentration?	Yes ☐ No ☐	If no, reduce cooling tower bleed rate to achieve higher levels of dissolved solids. Consult with your water treatment contractor for advice. Increasing concentration cycles can reduce water and chemical costs by 50%
	Are you considering replacing or retrofitting your water cooling tower?	Yes ☐ No ☐	If yes, consider installing an air-cooled chilled water system. Air-cooled units of less than 500kWR can be more cost effective and eliminate health risks
Cleaning	Is wet carpet cleaning methods used?	Yes ☐ No ☐	If yes, switch from steam carpet cleaning to dry powder cleaning
	Are windows cleaned regularly?	Yes ☐ No ☐	If yes, switch from a 'regular' to an 'as required' cleaning schedule
	Do you hose down bathroom floors?	Yes ☐ No ☐	If yes, use a mop rather than a hose
	Have you reviewed your toilet cleaning practices?	Yes ☐ No ☐	If no, ensure that toilet cleaners use water efficiently. Unnecessary flushing is a common occurrence with cleaning staff
Kitchen	Are taps left running during food preparation?	Yes ☐ No ☐	If yes, install foot-operated faucets or sensor-controlled taps
	Are dishes rinsed prior to washing?	Yes ☐ No ☐	If yes, scrape dishes before washing
	Are dishes washed by hand?	Yes ☐ No ☐	If yes, don't let the water run while rinsing. Fill one sink with wash water and the other with rinse water
	Is food washed under running water?	Yes ☐ No ☐	If yes, wash fruit and vegetables in a basin of water
	Is running water used to defrost food?	Yes ☐ No ☐	If yes, plan ahead and defrost in the fridge overnight, or use a microwave
	Is more ice produced than required?	Yes ☐ No ☐	If yes, adjust ice machines to dispense less ice
	Do you use water-cooled ice machines?	Yes ☐ No ☐	If yes, where cost effective replace the water-cooled units with air-cooled ice machines
	Is the dishwasher loaded fully before use?	Yes ☐ No ☐	If no, wash full loads only. Otherwise, set the load adjustment button on your washing machine to actual load conditions

Area	Question	Answer	Action
	Do you hose down kitchen floors?	Yes ☐ No ☐	If yes, consider alternatives such as sweeping, using a mop or squeegee
	Are spray guns used when hosing is necessary?	Yes ☐ No ☐	If no, use trigger-operated spray guns on all hoses
Laundry	Do you wash full loads?	Yes ☐ No ☐	If no, wash full loads only
	Have you evaluated the machine cycle efficiency?	Yes ☐ No ☐	If no, consult with your supplier to determine if it is possible to eliminate a cycle from the wash
	Do you reuse the water from the final rinse?	Yes ☐ No ☐	If no, consider installing a storage tank to recover water from the final rinse cycle for use in the next load, which can reduce water usage by 15%
	Are front loading washing machines utilised?	Yes ☐ No ☐	If no, consider using front loading washing machines, which use only half the amount of water used in conventional top loading washers. Front loading washing machines will also save energy
Monitoring	Is at least one of your staff responsible for water usage?	Yes ☐ No ☐	If no, assign a water-efficiency manager and link water conservation outcomes to performance review
	Is site water usage regularly monitored?	Yes ☐ No ☐	If no, read water meter(s) at least once a week. Graph data to identify usage trends. This will also help identify leaks as they occur
	Do you have a permanent water monitoring system?	Yes ☐ No ☐	If no, consider installing a permanent monitoring system to assist in the identification of leaks and other problems
	Are staff aware of trends in water usage?	Yes ☐ No ☐	If no, communicate water usage trends and progress reports through staff bulletins, newsletters and emails
	Do you benchmark your water usage?	Yes ☐ No ☐	If no, benchmark water usage against best practice. If below industry standards then carry out an audit to identify water-saving opportunities
Maintenance	Are leaks checked for in your regular maintenance schedule?	Yes ☐ No ☐	If no, check the water supply for leaks and turn off unnecessary flows during maintenance checks. Report leaks and other water losses immediately. Dye tests can help identify leaks
	Are there unused areas or equipment within your facility?	Yes ☐ No ☐	If yes, shut off the water supply to areas and equipment that are not used
	Is water efficiency formally considered when replacing fixtures?	Yes ☐ No ☐	If no, ensure low flow/high efficiency devices are installed when replacing water-using equipment. AAA-rated dishwashers, for example, use 18L of water per load, whereas older units use up to 60L per load
	Are overflow pipes regularly checked?	Yes ☐ No ☐	If no, ensure all overflow pipes on tanks, cooling towers and other equipment are checked on a weekly basis
Outdoor[1]	Is the garden bed mulched?	Yes ☐ No ☐	If no, use mulch to prevent weed growth and reduce evaporation loss by 70%
	Is the garden regularly watered?	Yes ☐ No ☐	If yes, look for signs of wilt before watering established plants. Also, water early morning or late evening to minimise moisture loss
	Does your garden have drought-tolerant plants?	Yes ☐ No ☐	If no, adopt drought-tolerant, native plants and lawns like Nioka and Palmetto which require less water and are low maintenance
	Do you have a drip irrigation system?	Yes ☐ No ☐	If no, consider installing a drip irrigation system, which is the most efficient method of irrigating and can save up to 50% over conventional systems
	Do you separately meter your irrigation supply?	Yes ☐ No ☐	If no, separately meter water supply so that usage can be monitored and adjusted to prevent over watering
	Do you hose down hard surfaces?	Yes ☐ No ☐	If yes, sweep with a broom or use a blower instead to clean sidewalks and driveways
Pools	Is splash-out common?	Yes ☐ No ☐	If yes, lower pool levels
	Do you back-flush filters daily?	Yes ☐ No ☐	If yes, consider reducing back-flush frequency to once a week in summer and once a fortnight in winter
	Does the pool area have a wash bay?	Yes ☐ No ☐	If yes, retrofit shower with water-efficient shower head rated at 9L/min
	Is your pool maintained by external contractors?	Yes ☐ No ☐	If yes, consider managing your pool in-house. External contractors tend not to focus on water usage and are also located off site. Problems encountered between scheduled visits can go unchecked for several weeks
	Is your pool area hosed down frequently?	Yes ☐ No ☐	If yes, consider reducing hose down frequency or using a mop. Install water-efficient spray guns on your hoses
	Is your pool heated?	Yes ☐ No ☐	If yes, reduce water temperature if possible to minimise water evaporation. Energy savings can also be substantial
	Do you use pool covers?	Yes ☐ No ☐	If no, install a pool cover to reduce evaporation and heat loss when the pool is not used. Without a cover, over half of the water in your pool can evaporate over a year
Training	Are awareness campaigns conducted amongst staff and customers?	Yes ☐ No ☐	If no, consider using signage to convey water conservation messages. Posters, hotel room cards and maintenance stickers are available from Sydney Water
	Are staff trained how to operate water-using equipment?	Yes ☐ No ☐	If no, train staff about the efficient operation of water-using equipment. This information should be readily available form the vendor
	Do you encourage ideas from staff to save water?	Yes ☐ No ☐	If no, seek employee suggestions through regular team meetings and create a water conservation ideas box for innovative ideas

[*] For more information on cooling tower operation refer to Sydney Water's 'Best practice guidelines for cooling towers in commercial buildings.' This document is available on the Sydney Water's website; www.sydneywater.com.au

[†] Mandatory water restrictions are now in force across Sydney, Illawarra and the Blue Mountains. Check Sydney Water's website for details; www.sydneywater.com.au

Conversions

SI Prefixes

Prefix name	Prefix symbol	Power-of-ten
yocto	y	10^{-24}
zepto	z	10^{-21}
atto	a	10^{-18}
femto	f	10^{-15}
pico	p	10^{-12}
nano	n	10^{-9}
micro	μ	10^{-6}
milli	m	10^{-3}
centi	c	10^{-2}
deci	d	10^{-1}
deka	da	10^{+1}
hecto	h	10^{+2}
kilo	k	10^{+3}
mega	M	10^{+6}
giga	G	10^{+9}
tera	T	10^{+12}
peta	P	10^{+15}
exa	E	10^{+18}
zeta	Z	10^{+21}
yotta	Y	10^{+24}

Length

Metre (m)	Millimetres (mm)	Inches (in.)	Feet (ft)
1	1000	39.37	3.28
0.001	**1**	0.039	0.0032
0.0254	25.4	**1**	0.083
0.3048	304.80	12	**1**

1 micrometre (μm) $= 1 \times 10^{-3}$ mm $= 4 \times 10^{-5}$ in.
1 thousandth of an inch (mil) $= 25.4$ μm
1 kilometre $= 1000$ m $= 0.621$ miles

Area

Square metre (m^2)	Square (ft^2)	Hectare (ha)	Acre (a)
1	10.764	0.0001	0.00247
0.093	1	9.29×10^{-6}	2.29×10^{-5}
10,000	107,640	1	2.471
4,046	43,561	0.4047	1

1 square kilometre (km^2) = 100 ha = 0.38610 square miles

Volume

Liters (L)	Gal. (US) (US gal.)	Gal. (UK) (Imp. gal.)	Cubic meters (m^3) (kL)	Cubic feet (ft^3)
1	0.264	0.220	0.001	0.035
3.785	1	0.833	0.00378	0.134
4.546	1.20	1	0.004546	0.160
1000	264	222	1	35.315
28.32	7.481	6.23	0.02832	1

1 acre-foot	= 325 851 US gal. = 1234 m^3
1 barrel	= 159 L = 42 US gal.
1 ton of crude oil	= approximately 7.3 barrels
1 quart (US)	= 0.964 L
1 quart (UK)	= 1.136 L

Flow

Litres/second (L/s)	Gal. (US)/minute (US gpm)	Gal. (UK)/minute (Imp. gpm)	Cubic meters/day (m^3/day) (kL/day)	Cubic feet/second (ft^3/s)
1	15.9	13.2	86.4	0.035
0.06	1	0.83	5.44	0.002
0.08	1.20	1	6.54	0.003
0.012	0.18	0.15	1	0.0004
0.008	0.12	0.10	0.68	1

1 acre -ft/day	= 14.276 L/s
1 million US gal./day (mgd)	= 3.069 acre-ft/day

Weight

Kilogram (kg)	Pound (lb)	Ton (ton)
1	2.20	0.001
0.456	**1**	4.55×10^{-4}
1,000	2,200	**1**

1 kilogram (kg) = 1000 grams
1 kilogram (kg) = 35.274 ounce
1 ton   = 0.984 long ton
1 ton   = 1.1 short ton
1 gram   = 15.4 grains

Pressure

kiloPascal (kPa)	Pound per square inch (psi)	Bar
1	0.14505	0.01
6.894	**1**	0.00689
100	14.5034	**1**

1 kiloPascal (kPa) = $0.0102 \, kg/cm^2$
1 kiloPascal (kPa) = 2.985 ft of water (ft-H_2O)
1 kiloPascal (kPa) = 7.5 torr
1 mm Hg   = 0.133 kPa
1 in. Hg   = 3.386 kPa

Energy

Kilo Joule (kJ)	British Thermal Unit (Btu)	kilo Calorie (kcal)	kWh
1	0.9478	0.238	2.78×10^{-4}
1.055	**1**	0.252	2.93×10^{-4}
4.1868	3.968	**1**	1.16×10^{-3}
3,600	3,412.3	859.8	**1**

1 therm (US) = 100 000 Btu = 105.5 MJ.

Power

Kilo Watt (kW)	British Thermal Unit per hour (Btu/hr)	Horsepower (UK) (hp)
1	3,412	1.342
2.93×10^{-4}	**1**	3.93×10^{-4}
0.7457	2.54×10^3	**1**

1 MW   = 3.6 GJ
1 ton refrigeration   = 3.517 kW
1 boiler horsepower = 9.810 kW

Index

Printed and bound by CPI Group (UK) Ltd, Croydon, CR0 4YY

03/10/2024

01040436-0018